Science, the State, and the City

Science, the State, and the City

Britain's Struggle to Succeed in Biotechnology

Geoffrey Owen and Michael M. Hopkins

OXFORD

UNIVERSITY PRESS

OXFORD

UNIVERSITY PRESS

Great Clarendon Street, Oxford, OX2 6DP,
United Kingdom

Oxford University Press is a department of the University of Oxford.
It furthers the University's objective of excellence in research, scholarship,
and education by publishing worldwide. Oxford is a registered trade mark of
Oxford University Press in the UK and in certain other countries

First Edition published in 2016
Impression: 1

Published in the United States of America by Oxford University Press
198 Madison Avenue, New York, NY 10016, United States of America

British Library Cataloguing in Publication Data
Data available

Library of Congress Control Number: 2015949485

ISBN 978–0–19–872800–9

Printed in Great Britain by
Clays Ltd, St Ives plc

Acknowledgements

This book is the result of a collaboration between two authors both of whom had long been seeking to publish a history of the emergence of the UK biotech sector. A chance meeting at a conference organized by Mariana Mazzucato in February 2012 led to a long correspondence and ultimately gave rise to this joint effort.

Geoffrey Owen led a programme of more than eighty interviews with current and former biotech executives, academic scientists, investors, venture capitalists, and government officials in the UK, Continental Europe, and the US. To them and all the others with whom we talked we are grateful for their patience in answering our questions and often engaging in further dialogue after the interview.

The US chapter could not have been written without the assistance of Nigel Gaymond, who used his unrivalled knowledge of the US biotech scene to arrange interviews for us in Boston and to put us in touch with British expatriates working in US biotech firms. On Japan, we benefited from extensive conversations with Robert Kneller at the University of Tokyo. We are also grateful to Thomas Heimann of HBM Partners for guidance on the Swiss biotech sector, to Siegfried Bialojan of Ernst & Young and Georg Kääb from the Bavarian Biotechnology Cluster on Germany, and to Catherine Moukheibir of Innate Pharma on France.

Michael Hopkins came to this book project having already benefited from time researching the financing of the sector on a grant funded by the Engineering and Physical Sciences Research Council (EP/E037208/1), research led by Charles Baden-Fuller (Cass Business School) and Paul Nightingale (SPRU, the Science Policy Research Unit, at the University of Sussex). We drew greatly upon insights and materials gathered during that project. These provided an essential first step in tackling the challenge of a broader sectoral history. For help in marshalling data and producing charts and figures, we are grateful for the efforts of Philippa Crane, Samuel Grange, and Thomas Welford. Scholarly advice and encouragement has come from Paul Nightingale, Ed Steinmueller, and Johan Schot at SPRU.

We owe a particular debt to Tera Allas, William Bains, John Barber, Kean Birch, Sir David Cooksey, Samir Devani, Daniel Green, Susan Hopkins,

Andrew Jack, Andrew Joy, Daniel Mahony, and Jack Scannell, who read and commented on draft sections of the book.

Michael Hopkins thanks Sam, Ben, and Emily for their patience, support, and forbearance.

Geoffrey Owen thanks Miriam for extensive editing and helpful comments throughout.

Contents

List of Figures

List of Tables

1

A New Way of Making Medicines

Biotechnology is the next wave of the knowledge economy, and I want Britain to become its European hub. . . . The science of biotechnology is likely to be, to the first half of the 21st century, what the computer was to the second half of the 20th century. Tony Blair, British Prime Minister, November 2000.[1]

When James Watson and Francis Crick published the double helix structure of DNA in 1953, their discovery was hailed as a great scientific achievement, but it had no obvious or immediate commercial implications. Twenty-five years later, as other scientists working in the field of molecular biology and genetics built on what Watson and Crick had done, a new technology emerged which came to be seen as comparable in importance to the computer. This was biotechnology, a set of techniques through which living organisms could be manipulated or modified to create useful products. When the commercial potential of these techniques became clear, they attracted intense interest from governments and investors throughout the world. Governments saw biotechnology as a means of creating new industries and reshaping old ones, investors as a means of making money.[2]

The new techniques have affected many industries, including agriculture, where the pros and cons of genetically modified crops are hotly debated, but their biggest impact has been in medicine. Biotechnology opened up new approaches for drug discovery and manufacturing and, over time, enabled scientists to launch novel treatments for intractable diseases such as cancer and multiple sclerosis. These techniques were radically different from the research methods on which the pharmaceutical industry had relied for many years. Partly for that reason, the exploitation of biotechnology in medicine, which began in the 1970s and 1980s, was led by newly formed

---

[1] Speech by the British Prime Minister at the European Bioscience Conference, 17 November 2000.

[2] The role of governments in the early years of biotechnology is described in Luigi Orsenigo, *The Emergence of Biotechnology* (London: Pinter, 1989) ch. 6.

entrepreneurial firms, many of them founded or co-founded by academic scientists. The established pharmaceutical companies, with some exceptions, were slow to see the relevance of biotechnology to their business, and left the field open to new entrants.

There are some parallels between biotechnology and the commercialization of semiconductors after the invention of the transistor in 1948. This invention, which built on many years of academic research in solid state physics, was made by scientists working in a large industrial company, but the subsequent application of the technology was driven by new firms such as Intel, not by the established manufacturers of electronic components.[3] Semiconductors gave birth to the personal computer and here, too, as in other branches of what came to be called the information technology industry, much of the dynamism came from new firms.

Many of the pioneers in information technology were based in the US, some of them in Silicon Valley, which became a hotbed of innovation in a range of electronic devices and equipment. Several of these US companies, such as Intel, Apple, and later Google, have built immensely powerful businesses; only a few non-American firms, such as Samsung in South Korea, have caught up with the American leaders in some branches of the industry.

US dominance is even more marked in biotechnology. The first biotechnology firms were founded in the US, and the US biotechnology sector has been responsible, directly or indirectly, for a large proportion of the innovative medicines that have been launched in recent years.[4] The two great regional biotechnology clusters around Boston and San Francisco have spawned a continuing stream of new firms, mostly linked to universities, several of which have achieved multi-billion dollar valuations in the stock market. Other countries have tried to build similar clusters, just as they have tried to replicate Silicon Valley in information technology, but with limited success.[5]

Innovation Ecosystems and Industrial Success

This book describes how the business of biotechnology has evolved since the 1970s, and examines why it has been so difficult for firms in other countries to

[3] Richard R. Nelson, 'The link between science and invention: the case of the transistor', in *The Rate and Direction of Inventive Activity: Economic and Social Factors*, A report of the National Bureau of Economic Research, New York (Princeton: Princeton University Press, 1962); John Tilton, *International Diffusion of Technology: The Case of Semiconductors* (Washington, DC: Brookings Institution, 1971).

[4] Robert Kneller, 'The importance of new companies for drug discovery: origins of a decade of new drugs', *Nature Reviews Drug Discovery* 9 (November 2010) pp. 867–82.

[5] Steven Casper, *Creating Silicon Valley in Europe, Public Policy Towards new Technology Industries* (Oxford: OUP, 2007).

emulate the success of American entrepreneurs in building world-leading biotechnology businesses and producing therapeutically important drugs. In several countries the quality of academic science was on a par with that of the US, but the US had other advantages which, taken together, constituted an extraordinarily dynamic 'innovation ecosystem'.

We use the term ecosystem to refer to the context within which innovations in a particular national industry occur, and to highlight the importance of interactions between different types of organizations with specialist functions and expertise. Within an innovation ecosystem the activities of all actors—government, firms, and individuals—are guided by institutions, policies, cultural norms, and traditions which influence the way that innovation processes are directed, resourced, and conducted.[6]

To illustrate the differences between the US and follower countries the book looks in detail at the UK and in particular at how successive British governments from 1980 onwards sought to promote biotechnology and to create institutions similar to those that had underpinned US success.[7] The book then compares these policies with the steps taken in other countries to improve the environment for biotechnology firms. The comparison makes it clear that, while there have been some successes in other countries, the gap between the US and the rest of the world in biotechnology has widened rather than narrowed in recent years.

Beyond Biotechnology

While this book is about biotechnology, it also sheds light on the wider question of why new industries, and science-based industries in particular, flourish in some countries more than in others. Many governments believe that these industries make an important contribution to economic growth, and, despite uncertainty about their prospects, need to be encouraged. The challenge is to put in place a set of institutions and policies that support

---

[6] The approach used in this book has its roots in the study of the economics of innovation. In this literature the term 'innovation ecosystem' has been used recently, as a variation of the more established term 'national system of innovation'. The term innovation ecosystem implies a set of inter-connected organizations, perhaps with very different aims, working within a common business environment such as a nation or region. B. -Å. Lundvall (ed), *National Innovation Systems: Towards a Theory of Innovation and Interactive Learning* (London: Pinter, 1992); Richard R. Nelson (ed.), *National Systems of Innovation, a Comparative Analysis* (Oxford: OUP, 1993); Mariana Mazzucato, *The Entrepreneurial State: Debunking Public vs. Private Sector Myths* (London: Anthem Press, 2013).

[7] Following Nelson, the term 'institutions' as used in this book refers to laws, widely held rules or traditions that guide behaviour within a particular innovation ecosystem. See Richard R. Nelson, 'What enables rapid economic progress: what are the needed institutions?' *Research Policy* 37 (2008) pp. 1–11.

scientific research and promote its commercialization. In some technologies, of which biotechnology is one, the American ecosystem has worked outstandingly well in terms of generating large, commercially successful firms and innovative products and services.

In contrast to industries such as the automotive industry, where American manufacturers have lost ground to Japanese firms such as Toyota and Honda, US leadership in biotechnology has not so far been seriously challenged. By setting out the events which followed the emergence of biotechnology in the 1970s, this book seeks to identify the sources of US success and the obstacles which prevented other countries from matching it.

The Scientific Breakthroughs behind Biotechnology

Historians use the term 'old biotechnology' to refer to the use of fermentation to produce alcohol and other food products; these techniques were well established in the nineteenth century.[8] It was also possible for commercially useful molecules, such as vitamins, to be extracted through the fermentation of micro-organisms and used for therapeutic purposes.

The distinguishing feature of 'new biotechnology' is the ability to transfer useful characteristics encoded by genes from one species to another and to harness the productive capacities of natural cells to produce a wider range of valuable molecules at commercial scale than has previously been possible. Biotechnology in its modern form was built on scientific advances that began in the 1940s when Oswald Avery and his colleagues at the Rockefeller Institute, New York, identified Deoxyribose Nucleic Acid (DNA) as the carrier of genetic information, capable of transferring characteristics from generation to generation and from organism to organism. In the early 1950s Watson and Crick in Cambridge, with support from Maurice Wilkins and Rosalind Franklin at Kings College London, worked out the structure of DNA and its mechanism of replication.

Over the next decade teams of scientists in France, the UK, and the US worked to crack the genetic code, showing how the information encoded as a sequence of nucleic acids on a strand of DNA could be translated into instructions to make a given protein. From the late 1960s onwards hundreds of restriction enzymes (proteins that recognize specific, short nucleotide sequences) were identified that could act as molecular scissors to cut strands of DNA.

---

[8] One of the earliest uses of the word 'biotechnology', at the time of the First World War, was by a Hungarian agricultural engineer, Karl Ereky, to refer to the process whereby raw materials could be biologically upgraded. Robert Bud, *The Uses of Life, a History of Biotechnology* (Cambridge: CUP, 1993) pp. 32–5.

Through the use of these enzymes, further experiments showed that genetic characteristics could be excised from one organism and transferred or 'cloned' between species. In 1973 Stanley Cohen at Stanford University and Herbert Boyer at the University of California, San Francisco, developed what became known as the recombinant DNA technique for creating genetically modified organisms. These organisms became the basis for the industrial production of complex biological molecules for a host of applications. During the 1980s several drugs were produced using recombinant DNA technology and launched with commercial success, including insulin and human growth hormone.

While recombinant DNA was invented in the US, a separate line of research was under way in the UK. In 1976 César Milstein and Georges Köhler, working at the Medical Research Council's Laboratory of Molecular Biology in Cambridge, found a way of making monoclonal antibodies. These are proteins which recognize and attach to specific molecules, marking them for destruction by the body's immune system.[9] The impact of this technology for therapeutic applications was not as immediate as that of recombinant DNA. The first monoclonal antibody drugs were derived partly from mice and produced adverse immune reactions in patients. During the 1980s and 1990s scientists worked to address these problems by developing humanized and later fully human monoclonal antibodies.[10] These technologies are used in many of today's best selling drugs, including Humira (adalimumab) for rheumatoid arthritis and Herceptin (trastuzumab) for breast cancer.[11]

Recombinant DNA and monoclonal antibodies were the key technologies which underpinned the growth of the biotechnology sector. Another strand was the development by American and British scientists, including Fred Sanger in Cambridge, of techniques that made it possible to read the sequence of nucleic acids on a strand of DNA. Scientists were able to read long genetic sequences and to detect even the smallest changes; this paved the way for the mapping of the human genome in the 1990s and early 2000s. Better understanding of the form and function of genetic sequences also opened up possibilities for new treatments such as gene therapy, through which genes are inserted into a patient's cells to replace faulty genes. Technical advances brought sharp falls in the cost of genetic analysis, and the linking of these techniques with the increasing power of computers made genomics a

---

[9] The usefulness of monoclonal antibodies (MAbs) stems from the specificity with which they bind to their target antigen and the ability to develop MAbs that can bind effectively to a wide range of molecular targets. They have applications as diagnostics, as markers for the presence of their target molecules, and as therapeutics, as they can bind to molecules and cells, inhibiting their function and marking them for destruction by the cells of the immune system.

[10] Lara V. Marks, *The Lock and Key of Medicine: Monoclonal Antibodies and the Transformation of Healthcare* (New Haven: Yale University Press, 2015).

[11] In referring to new drugs, this book gives the trade name (e.g. Humira) and the generic name. Different trade names may be used in different geographies.

powerful research tool for guiding drug discovery; genomics became a subfield within biotechnology.

The Pharmaceutical Industry and Biotechnology

Advances in molecular biology and genetics were at the heart of the emerging biotechnology sector. By contrast, the pharmaceutical industry, a much older industry made up of large, well-established companies, relied on chemistry rather than biology as the principal route to drug discovery. As long as these methods were generating a continuing flow of profitable drugs, there seemed to be no obvious reason to change them.

In the eighteenth and nineteenth centuries most medicines, such as quinine for malaria and other fevers, were derived from botanical sources. In the second half of the nineteenth century the industry entered a new phase as a result of scientific research, principally in Germany, in the field of synthetic organic chemistry. Building on the work of Justus von Liebig and other German university scientists, companies such as Hoechst and Bayer developed synthetic dyestuffs, replacing dyes derived from plants.[12] They later found that the same technology could be used to produce drugs; the invention of aspirin, launched by Bayer in 1899, is a celebrated example. German firms, together with a group of Swiss manufacturers that included Ciba, Geigy, and Sandoz, dominated the supply of chemistry-based drugs up to the Second World War.[13]

Another theme in the history of the pharmaceutical industry is microbiology, pioneered by Robert Koch in Germany and Louis Pasteur in France in the late nineteenth century, leading to a greater understanding of the bacterial origins of infectious diseases. A major breakthrough was penicillin, discovered in the UK in 1928 and fully developed by British and American scientists during the Second World War. The success of penicillin set off a wave of innovation in antibiotics, and encouraged drug manufacturers to invest much more in research than they had done before the war.

The period between the 1950s and the 1980s is often referred to as a golden age for the pharmaceutical industry, as a stream of new drugs was introduced to treat hitherto incurable diseases. During these years, thanks largely to the antibiotics, industry leadership passed from Germany and Switzerland to the US. Companies such as Merck and Pfizer built large and highly productive

---

[12] Johann Peter Murmann, *Knowledge and Competitive Advantage* (Cambridge: CUP, 2003).
[13] Basil Achilladelis, 'Innovation in the pharmaceutical industry', in *Pharmaceutical Innovation, Revolutionising Human Health*, edited by Ralph Landau, Basil Achilladelis, and Alexander Scriabine (Philadelphia: Chemical Heritage Press 1999).

research laboratories, supported by a large sales force whose task was to persuade doctors to prescribe their medicines. A group of British firms, led by Glaxo, adopted a similar approach and also achieved considerable success.[14]

There were improvements in research techniques during this period, thanks to advances in pharmacology, medicinal chemistry, and other disciplines. Random screening—testing a wide range of natural or synthetic chemical compounds in the laboratory and in animals to see whether they had therapeutic properties—gave way to 'drug discovery by design', based on a more complete understanding of the causes of disease.[15] But the output of the industry continued to consist largely of small molecule drugs which were produced through chemical synthesis and had a well-defined chemical structure; most of them were supplied in tablet or liquid form.

This was an entirely different approach from that followed by the new biotechnology firms. Their manufacturing techniques involved the reconfiguring of living cells to produce protein-based large molecule drugs (generally known as biologics) that would be impractical or even impossible to make through the chemical processes on which the pharmaceutical companies relied. Because of their protein composition, these drugs have to be injected or administered by infusion rather than swallowed. Whereas small molecule drugs are absorbed into the bloodstream and can reach almost any destination, large molecule drugs cannot be taken orally because they would be digested in the stomach or intestines and rendered ineffective.

The initial reaction to the emergence of biotechnology on the part of the established pharmaceutical companies—the largest of which are generally known as Big Pharma—was sceptical. The new techniques called for knowledge and capabilities which they did not possess, and their existing research and manufacturing methods were serving them well. During the 1980s, as Genentech, Amgen, and other US biotechnology firms brought biological drugs to the market, the pharmaceutical companies began to recognize that biotechnology had opened up a potentially fruitful approach to drug discovery; they also saw that biotechnology-based techniques could be used as tools to make the development of small molecule drugs more efficient.

What then ensued was a series of moves by the pharmaceutical companies to integrate the new techniques into their own operations, either through in-house research or by acquiring or partnering with biotechnology firms.

---

[14] By the 1970s Glaxo had emerged as the largest of the four leading British pharmaceutical manufacturers. The others were Wellcome (formerly Burroughs Wellcome), Beecham, and Imperial Chemical Industries.

[15] Rebecca Henderson, Luigi Orsenigo, and Gary P. Pisano, 'The pharmaceutical industry and the revolution in molecular biology', in *Sources of Industrial Leadership*, edited by David Mowery and Richard R. Nelson (Cambridge: CUP, 1999).

This led to a blurring of the borderline between 'pharma' and 'biotech' (the shortened form of 'biotechnology' was by now in common use), but it did not result in the domination by Big Pharma companies of large molecule as well as small molecule drug production, nor did it cause any diminution in the flow of new entrepreneurial firms eager to follow in the footsteps of the biotechnology pioneers. The biotech sector, as defined in the next section, remains a prolific source of innovative drugs.

Definitions and Scope

When the first biotechnology firms were created in the US in the 1970s, their defining characteristic was the use of recombinant DNA or monoclonal antibody technology. Over the next few years other drug discovery firms were established, in the US and in other countries, some of which built their business on small molecule drugs, utilizing technologies similar to those long used by the established pharmaceutical companies. These firms were not 'pure' biotechnology firms, but they were generally classified by commentators and analysts as biotechs; it suited their founders to be seen as part of the new wave of drug discovery which was attracting considerable interest from investors. With the pharmaceutical companies moving strongly into biologics, the question 'what is a biotech firm?' became harder to answer.[16]

To complicate the definitional problems further, some drug discovery firms came to be classified as speciality pharmaceutical companies. These are generally smaller than Big Pharma and concentrate on developing or acquiring drugs for niche markets including so-called orphan diseases, such as cystic fibrosis or muscular dystrophy. Because the patient populations for such drugs are relatively small, they have generally not attracted the attention of the large pharmaceutical companies. Orphan drugs are marketed to medical specialists rather than doctors in general practice, and so do not require a large sales force. Some speciality pharma companies, such as Shire in the UK, were founded in the 1980s as part of the wave of drug discovery firms that were being formed at that time; others were in existence before the emergence of modern biotechnology.

---

[16] Differences between studies of the biotech sector in the definitions used and data collection methods make the assessment of long-term trends in the financing and productivity of national industries problematic. For this reason, although the book makes use of many prior studies on biotechnology it does not integrate data from these to make comprehensive direct comparisons. For discussion of the issues around such comparisons see: Michael M. Hopkins and Joshua Siepel, 'Just how difficult can it be counting up R&D funding for emerging technologies (and is tech mining with proxy measures going to be any better?)', *Technology Analysis and Strategic Management* 25, 6 (2013) pp. 655–85.

In studying the growth of the biotech sector, this book covers emerging drug discovery firms in general (the term 'therapeutic biotech firms' is used interchangeably), recognizing that despite technological differences their needs are similar and their fortunes linked. These firms discover and develop drugs that require the approval of regulators before they can be prescribed to patients. All of them share the challenge of bringing drugs through a heavily regulated development process as set out in Table 1.1. (The regulatory process is broadly similar in the US and in the other countries discussed in later chapters.)

The vast majority of drugs entering development fail, and even successful products generally take 10–14 years to move from discovery to market launch.[17] All firms engaged in bringing new drugs to market are judged against the progress of their products through this process, and its phases are frequently referred to throughout the book.

To distinguish between long established pharmaceutical firms and the new entrants, we use the generally accepted distinction between Big Pharma and biotech, where the term 'biotech' refers to any drug discovery firm founded after the key breakthroughs of the 1970s, irrespective of the technology it is using. In the UK, the starting point for the emergence of the first biotech firms is taken as 1980, for reasons explored in Chapter 3. The distinction between Big Pharma and biotech is not precise, and there is room for argument over, for example, whether Amgen, which makes both small molecule and large molecule drugs and is now as large as some pharmaceutical companies, should be regarded as big biotech or Big Pharma.

The book largely excludes (except where they are relevant to the main theme) companies which supply tools, equipment, or technology to biotech firms but are not themselves engaged in drug discovery. It does not discuss manufacturers of diagnostic equipment or medical devices (which form part of the wider 'life sciences industry'), nor does it venture into non-medical areas of application, such as agriculture.

Structure of the Book

Chapter 2 focuses on the US. It describes the progress of US biotech firms from the foundation of Genentech in 1976 to the most recent surge in biotech initial public offerings (IPOs) that began in 2013 and continued into 2015.[18] It highlights the constant flow of new entrants, the emergence of dynamic biotech clusters

---

[17] F. Pammolli, L. Magazzini, and M. Riccaboni, 'The productivity crisis in pharmaceutical R&D', *Nature Reviews Drug Discovery* 10 (June 2011) pp. 428–38.
[18] An Initial Public Offering (IPO) is the first occasion when a company's shares are made available for members of the public to buy and sell on a stock market. IPOs are usually used as an occasion to raise new capital from investors to support the development of the business.

Table 1.1. Key stages in the drug discovery and development process

Discovery	Pre-clinical development	Phase I clinical trials	Phase II clinical trials	Phase III clinical trials	Regulatory approval	Phase IV post marketing studies
Drug candidates are designed, screened and selected in order to target specific molecules in the body that are associated with disease.	Drug candidates are tested under laboratory conditions to identify evidence of effectiveness and toxicity in tissue or animals.	Candidate drugs are tested for safety in small numbers of individuals (often healthy volunteers).	Candidate drugs are tested for efficacy against a disease in patients. Often several hundred patients will be involved, including those given placebo treatments for comparative purposes.	Larger, longer trials will test candidate drugs for efficacy in the treatment of disease in patients. Sometimes several thousand patients will be involved, including those given placebo treatments.	Data from clinical trials are submitted to regulatory agencies responsible for licensing drugs for therapeutic use in humans. Following approval, drugs can be launched on the market.	After regulatory approval and market launch, further studies into the safety of a drug and its performance may be required by a regulator, or undertaken to demonstrate its advantages.

Source: Based on descriptions by the US Food and Drug Administration (www.FDA.gov) and Association of the British Pharmaceutical Industry (www.ABPI.org.uk).

around Boston and San Francisco, and the changing relationship between biotech firms and the pharmaceutical industry.

The next three chapters provide a history of UK biotechnology from 1980 to 2015. Chapter 3 describes the creation of Celltech, the UK's first dedicated biotech firm, and the subsequent evolution of the sector, including the opening of the London Stock Exchange to biotech firms in the early 1990s. Chapter 4 looks at the consequences of the technology bubble in 2000 and its sudden collapse, events which led to a more difficult financing environment for all biotech firms. A partial recovery in 2004–06 was followed by the world financial crisis of 2008–09, plunging the sector into a mood of pessimism about its future. Chapter 5 examines the extent of the recovery which took place between 2009 and 2015.

Chapter 6 reviews the attempts by three other European countries (Germany, France, and Switzerland) and by Japan to promote the growth of their biotech sectors, and assesses the effectiveness of the policy instruments that were used.

The following two chapters look at two issues which have been central to the debate about the performance of the UK biotech sector. Chapter 7 examines whether the growth of biotech firms in the UK has been held back by lack of support from the financial system. Chapter 8 explores the changing institutional context for British firms and assesses the impact of UK government policy.

Chapter 9 sums up the findings of the book, setting out why the UK and other countries have not been able to emulate the success of the early US biotechnology firms and assessing the prospects of this changing in the future.

2

The US Takes the Lead

It was like maybe a dam waiting to burst or an egg waiting to hatch, but the fact is, there were a lot of Nobel Prizes in molecular biology but no practical applications. Ronald Cape, co-founder of Cetus.[1]

The first US firms to exploit the scientific breakthroughs in biotechnology looked for commercial opportunities in several industries; it was not obvious at the start that the most lucrative opportunities would be in medicine. Those that focused on healthcare were entering an industry which was dominated by large, long-established pharmaceutical companies such as Merck, Pfizer, and Eli Lilly.

This was an industry which had acquired its modern form in the years following the Second World War.[2] Before the war very few diseases could be cured by drugs and the industry spent little on research. The advent of penicillin, followed by other antibiotics, opened up a new and highly profitable route to drug discovery. These 'wonder drugs' were the principal contributors to the pharmaceutical industry's profits in the early post-war decades; spending on research as a proportion of sales rose from 4 per cent to 8 per cent between 1950 and 1960.

A powerful incentive for investment in research was the patent system.[3] Penicillin itself was not patented, but for the antibiotics that came later the US Patent Office ruled that where drugs were derived from natural substances, both the product itself and the manufacturing process were patentable. Merck obtained a patent for streptomycin in 1948, and over the next few years

---

[1] Ronald Cape, 'Biotech pioneer and co-founder of Cetus', oral history conducted in 2003 by Sally Smith Hughes, Regional Oral History Office, The Bancroft Library, University of California-Berkeley, 2006.

[2] Peter Temin, 'Technology, regulation and market structure in the modern pharmaceutical industry', *Bell Journal of Economics* 10 (1979) pp. 429–46.

[3] A utility patent allows the owner to exclude others within a specific jurisdiction, such as a nation state, from commercially exploiting the invention it discloses to the public. Patent terms have varied by country and over time, but up to 17 years of exclusivity could be gained in the US during the twentieth century (recently this has increased to 20 years).

patented prescription drugs became the lifeblood of the industry; they were marketed principally to doctors and supported by heavy expenditure on advertising. In some cases a single drug accounted for the bulk of the company's sales and profits.

What emerged during the 1950s and 1960s was a group of about forty research-intensive, and advertising-intensive, companies.[4] Most of them were vertically integrated, handling all parts of the value chain from research through clinical development and production to marketing and distribution. Barriers to entry were high, but economies of scale were not so large as to cause the industry to become concentrated in the hands of a few giant companies; during these years there were few mergers among the leading firms (in contrast to what happened in the 1980s and 1990s).

Barriers to entry became even higher after the introduction of stringent regulations over drug safety in 1962.[5] The new rules made the process of obtaining the approval of new drugs more complex, and increased the costs of research. However, for firms that were able to navigate their way through the regulatory controls, the potential rewards were large. The US market for pharmaceuticals was not only much bigger than that of other countries, but also, in contrast to the UK and most other European countries, there were no government controls over drug prices. Most patients' medical bills were paid by private insurance companies or, from the mid-1960s, by the two government support programmes, Medicare and Medicaid.

Success in pharmaceuticals depended primarily on company-financed research and development, but the industry benefited from the high level of government support for basic scientific research in medicine and other health-related disciplines. This support was channelled through the National Institutes of Health and other agencies and was mostly conducted in universities and research institutes.[6] Government funding contributed to advances in knowledge about the causes of disease, as well as expanding the supply of well-trained scientists, many of whom were employed in the pharmaceutical industry; it was an important component of the US life sciences innovation system.[7]

---

[4] Ian M. Cockburn, 'The changing structure of the pharmaceutical industry', *Health Affairs* 23, 1 (2004) pp. 10–22. The post-war evolution of the industry is discussed in Basil Achilladelis, 'Innovation in the pharmaceutical industry', in *Pharmaceutical Innovation*, edited by Ralph Landau, Basil Achilladelis, and Alexander Scriabine (Philadelphia: Chemical Heritage Press, 1999).

[5] This was a response to the thalidomide tragedy in the late 1950s. Thalidomide was a drug used to relieve morning sickness in pregnant women, and caused severe birth defects in babies.

[6] David C. Mowery and Nathan Rosenberg, *Paths of Innovation, Technological Change in 20th Century America* (Cambridge: CUP, 1998) p. 97.

[7] Iain M. Cockburn and Scott Stern, 'Finding the endless frontier: lessons from the life sciences innovation system for technology policy', in *Capitalism and Society* 5, 1 (2010); Iain M. Cockburn, Rebecca Henderson, Luigi Orsenigo, and Gary P. Pisano, 'Pharmaceuticals and biotechnology', in *US Industry in 2000, Studies in Competitive Performance*, edited by David Mowery (Washington, DC: National Academy Press, 1999).

Table 2.1. World's leading pharmaceutical companies in 1970

Company	Nationality	Sales ($m)
Roche	Switzerland	840
Merck	US	670
Hoechst	Germany	497
Ciba-Geigy	Switzerland	492
American Home Products	US	479
Lilly	US	421
Sterling	US	418
Pfizer	US	416
Warner Lambert	US	408
Sandoz	Switzerland	346

Source: Monopolies Commission, a report on proposed mergers between Beecham and Glaxo and between Boots and Glaxo (HC341, HMSO, July 1972).

By the 1970s the US pharmaceutical industry was the largest in the world, with six American companies ranked among the top ten manufacturers (Table 2.1). The structure of the industry was stable, with the manufacturers competing against each other on the basis of research methods that had served them well and were continuing to generate healthy profits. At this point a new and potentially disruptive technology entered the scene.

The First Biotechnology Firms

The impetus for the creation of the new biotechnology firms came from collaboration between academic scientists and venture capitalists, reinforced in some cases by the recruitment of experienced managers from established companies. Several of the venture capitalists who invested in biotechnology in the 1970s and 1980s had previously made money in electronics. However, as became clear after the first flush of enthusiasm had passed, biotechnology differed in several respects from electronics. The time scale between initial idea and marketable product was generally longer, and the risk of failure in the course of development was higher. The links with academia were also much closer. In contrast to electronics companies like Intel or Apple, many of the early biotech firms were founded or co-founded by academics, some of whom retained their university posts while working part-time for the companies they had created. To a much greater extent than in other industries, building a biotechnology business called for a fusion between two different worlds, that of academic science and that of commerce.[8]

---

[8] John F. Padgett and Walter W. Powell (eds), *The Emergence of Organisations and Markets* (Princeton: Princeton University Press, 2012) p. 375.

The first firm created with a specific plan to exploit advances in molecular biology was Cetus, set up in San Francisco in 1972. The founders were Moshe Alafi, an Iraqi-born venture capitalist who had acquired a taste for business while a graduate student at Berkeley, and Donald Glaser, a Nobel Prize-winning physicist.[9] They joined forces with two venture capitalists, Ronald Cape and Peter Farley, both of whom had a scientific background; Cape had been a post-doctoral researcher in molecular biology at Berkeley, while Farley was a medical doctor.

Cetus's first project was the development of an instrument invented by Glaser for screening colonies of microorganisms for useful properties, but this venture was not successful. Cetus evolved from instrumentation into the development of drugs, including vaccines and antibiotics. Although it assembled a team of talented scientists, Cetus's research efforts were spread too widely and its commercial direction was poor.[10]

The most successful of the pioneers, often seen a role model for the rest of the sector, was Genentech. This firm was founded in San Francisco in 1976 by Robert Swanson, a venture capitalist, and Herbert Boyer, a scientist at the University of California, San Francisco (UCSF), who had been the co-inventor of recombinant DNA. Swanson had spent 5 years with Citibank before joining Kleiner Perkins, a leading San Francisco venture capital firm, in 1974. Kleiner Perkins was an investor in Cetus, but it was unhappy with the company's unfocused approach to product development. Swanson's first assignment was to help the management work out a more effective commercial strategy.[11]

When that effort failed Kleiner Perkins sold its stake in Cetus, but by then Swanson had become fascinated by biotechnology. On his own initiative, he sought a meeting with Boyer. The scientist's initial reaction was that it was too early to think of starting a commercial venture based on recombinant DNA, but after further discussion the two men worked out a business plan that was sufficiently attractive to persuade Kleiner Perkins to provide seed funding of $250,000; the venture capital firm owned a third of the new company, with the rest held by Swanson and Boyer.[12] One of the partners, Tom Perkins, became chairman, with Swanson, then aged 29, as chief executive.

Swanson's idea from the start was to use recombinant DNA technology to produce and sell drugs, but this would take time and money. In the meantime,

---

[9] Glaser had won the Nobel Prize for physics in 1960 for the invention of the bubble chamber, a device for detecting electrically charged particles. He later switched from physics to molecular biology.

[10] Eric J. Vettel, *Biotech: The Counter-cultural Origins of an Industry* (Philadelphia: University of Pennsylvania Press, 2006) pp. 186–215.

[11] Sally Smith Hughes, *Genentech, the Beginnings of Biotech* (Chicago: Chicago University Press, 2011) pp. 31–2.

[12] Tom Perkins, *Valley Boy: The Education of Tom Perkins* (New York: Gotham Books, 2007) pp. 117–25.

to generate revenue, he planned to negotiate partnerships with pharmaceutical companies which would use Genentech's technology to complement their own research. Insulin, a treatment for diabetes, was seen by Genentech, and by other biotechnology firms that were founded around the same time, as a promising candidate for the new cloning technology.[13] At that time insulin was extracted from the pancreases of pigs and cows and Eli Lilly, the principal producer, was concerned whether supplies from these sources would keep pace with the increase in the diabetic population; animal-derived insulin also caused allergic reactions in some patients. In 1978 Lilly signed a 20-year agreement with Genentech whereby it acquired worldwide rights to manufacture and market human insulin, using the young firm's technology.[14] This involved an initial payment of $500,000, milestone payments as the research progressed, and royalties if and when human insulin manufactured using genetically engineered cells reached the market.

This agreement put Genentech on a more solid financial footing. It also set the pattern for future relationships between biotech and Big Pharma; licensing deals, contract research, and other forms of collaboration became vital sources of finance for biotech firms. Over the next few years Genentech negotiated several other partnership agreements, including one with Roche, the Swiss pharmaceutical firm, to produce interferons—a family of signalling proteins seen as having great therapeutic promise for a range of conditions, including cancer. But Swanson never deviated from his view that Genentech could only capture the full value of its investment in research by producing and selling its own drugs.

As Genentech was getting into its stride, there were two potential roadblocks which might have held back the growth of the biotechnology sector. One was concern within the scientific community, and in the public at large, about the risks of genetic engineering—the fear that the cloning of genes could get out of control and cause an environmental disaster through the release of superbugs. The need for safeguards was generally accepted, and the form they should take was discussed at a conference of scientists at Asilomar in California in 1975. The outcome was a 16 month moratorium during which the National Institutes of Health worked out a set of guidelines for genetic engineering experiments.[15] Although there continued to be unease in Congress, and some restrictions were imposed in individual states (including California and Massachusetts, which were later to become important hosts for biotechnology firms), these rules did not significantly

---

[13] Nicolas Rasmussen, *Gene Jockeys, Life Science and the Rise of Biotech Enterprise* (Baltimore: John Hopkins University Press, 2014) p. 40.

[14] Hughes, *Genentech*, p. 94.

[15] David Dickson, *The New Politics of Science* (Chicago: Chicago University Press, 1988) pp. 243–55.

hold back genetic engineering research, some of which was conducted outside the US.[16] By 1980 anxiety about the risks of cloning was giving way to a sense of excitement among politicians, commentators, and investors about the potential of the new techniques to transform the treatment of disease. The age of 'biomania' was dawning.[17]

The other uncertainty was whether organisms created by genetic engineering could or should be patented. Some scientists, especially in Stanford, were dismayed by the university's decision to apply for a patent on the Boyer–Cohen process; they objected strongly to the principle of private ownership of a basic research technology to which many scientists had contributed and which had been funded by taxpayers.[18] There was also a lack of clear guidance about the patentability of genetically engineered lifeforms from the US Patent and Trademark Office. It was not until 1980 that the legal position was clarified when the US Supreme Court in the Chakrabarty *versus* Diamond case ruled that genetically modified organisms made by man were potentially patentable under existing statutes. Genentech was not directly involved in the case—Ananda Chakrabarty was a General Electric scientist who had developed a bacterium for treating oil spills—but it filed an *amicus curiae* brief,[19] warning that an adverse decision would hold back the commercial development of genetic engineering and damage the prospects of young biotechnology firms.[20]

The Chakrabarty ruling opened the way for the next stage in Genentech's development—becoming a public company. Venture capital firms like Kleiner Perkins were backed by investors, known as limited partners, through the creation of investment funds that had a fixed duration, normally ten years; the success of the fund depended on achieving profitable exits from at least some of their investments within that ten-year period. Tom Perkins believed that the most likely outcome was for Genentech to be sold to a pharmaceutical company. He approached Johnson & Johnson and Eli Lilly, but neither took up his suggestion.[21] Given the excitement surrounding biotechnology and the enthusiasm among investors for high-technology ventures (Apple, led by

---

[16] Biogen, founded in 1978 by a group of American and European scientists, had its headquarters in Geneva, partly because of opposition from the city council in Cambridge, Massachusetts, to genetic engineering research. The company later relocated to Cambridge.

[17] Robert Teitelman, *Gene Dreams: Wall Street, Academia, and the Rise of Biotechnology* (New York: Basic Books, 1989.

[18] Doogab Yi, 'Who owns what? Private ownership and the public interest in Recombinant DNA Technology in the 1970s', *Isis* 102, 3 (September 2011) pp. 446–74. Doogab Yi, *The Recombinant University: Genetic Engineering and the Emergence of Stanford Biotechnology* (Chicago: University of Chicago Press, 2015).

[19] An *amicus curiae* brief is a document prepared and submitted to a court by an interested party and which has a bearing on the case but which has not been provided by the litigants.

[20] Sally Smith Hughes, 'Making dollars out of DNA; the first major patent in biotechnology and the commercialisation of molecular biology 1974–1980', *Isis* 92, 3 (September 2001).

[21] Hughes, *Genentech*, pp. 139–64.

Table 2.2. Biotech's first generation: IPOs from 1980 to 1986

Company	Date of IPO	Gross amount raised ($m)
Genentech	10/80	35
Cetus	3/81	107
Genetic Systems	4/81	6
Ribi Immunochem	5/81	1.8
Genome Therapeutics	5/82	12.9
Centocor	12/82	21
Bio-Technology General	9/83	8.9
Scios	1/83	12
Immunex	3/83	16.5
Amgen	6/83	42.3
Biogen	6/83	57.5
Chiron	8/83	17
Immunomedics	11/83	2.5
Repligen	4/86	17.5
OSI	4/86	13.8
Cytogen	6/86	35.6
Xoma	6/86	32
Genzyme	6/86	28
ImClone	6/86	32
Genetics Institute	5/86	79

Source: Cynthia Robbins-Roth, *From Alchemy to IPO, the Business of Biotechnology* (Cambridge, MA: Perseus Publishing, 2000) p. 21.

Steve Jobs, was preparing to go public at that time), Perkins opted for a stock market flotation.[22]

When Genentech was listed on NASDAQ in October 1980 the share price rose from $35 to $89 within twenty minutes before closing the day at $71. It was one of the most spectacular initial public offerings (IPOs) in Wall Street history; Boyer and Swanson became instant millionaires. The Genentech IPO, as Perkins remarked later, 'established the idea that you could start a new biotechnology company, raise obscene amounts of money, hire good employees, sell stock to the public. Our competitors started doing all that'.[23]

There were thirty-nine flotations of firms seeking funds to exploit biotechnology between 1980 and 1983, then a pause for breath as investors began to look more critically at what they were buying, followed by a revival of interest in 1986 and 1987 which allowed several more firms to go public (Table 2.2). This was a foretaste of the volatility which would continue to characterize stock market attitudes towards the biotech sector. Figure 2.1, which shows the number of biotech IPOs and the amounts of money raised annually from 1982 to 2014, illustrates the peaks and troughs of biotech financing throughout the period.

[22] Apple, founded in 1976 with venture capital backing, went public in December 1980.
[23] Hughes, *Genentech*, p. 161.

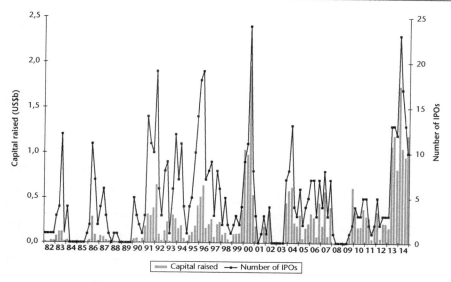

Figure 2.1. US biotech initial public offerings, 1982–2014
Source: Ernst & Young.

Most of the new firms, like Genentech, were founded by academic scientists and venture capitalists, with additional funding in some cases coming from industrial companies which saw biotechnology as relevant to their business or as a potentially profitable diversification. Three of the early investors in Biogen, a rival to Genentech co-founded by a Nobel Prize-winning Harvard scientist, Walter Gilbert, were International Nickel, a Canadian mining company, Schering-Plough, one of the few pharmaceutical companies to take an early and active interest in biotechnology, and Monsanto, which was converting itself from a traditional chemical company into a biotechnology-based business.

Investor Attitudes

As publicly listed companies, the new biotechnology firms had to manage their business in a way that was credible to investors. This meant, among other things, bringing drugs to the market quickly. Genentech's insulin was approved by the Food and Drug Administration in 1982, just 6 years after the company had been founded, and it was followed by a series of drug introductions by other firms, some of them involving partnerships with pharmaceutical companies (Table 2.3). Genentech itself went on to develop two drugs which it planned to market on its own, human growth hormone

Table 2.3. New biologics approved in the US, 1982–1992

Product (trade name)	Company (partner)	Indication(s)	Year
Insulin (Humulin)	Genentech (Eli Lilly)	Diabetes	1982
Human growth hormone (Protropin)	Genentech	Human growth hormone deficiency	1985
Interferon-alpha-2a (Pegasys)	Genentech (Roche)	Cancer, viral infections	1986
Interferon-alpha-2b (Intron)	Biogen (Schering-Plough)	Cancer	1986
Anti-T-Cell CD3 (Orthoclone OKT3)*	Ortho	Kidney transplant rejection	1986
Hepatitis B vaccine (Recombivax HB)	Chiron (Merck)	Hepatitis B	1986
Human growth hormone (Humatrope)	Eli Lilly	Human growth hormone deficiency	1987
TPA (Activase)	Genentech	Cardiovascular disease	1987
Erythropoietin (Epogen)	Amgen (Ortho)	Anaemia	1989
Interferon-alpha-n3 (Alferon)	Interferon Sciences	Genital warts	1989
Hepatitis B vaccine (Engerix-B)	Smith Kline Beecham	Hepatitis B	1989
Interferon-gamma (Actimmune)	Genentech	Cancer, infections, inflammatory diseases	1990
G-CSF (Neupogen)	Amgen	Chemotherapy effects	1991
GM-CSF (Prokine)	Immunex (Hoechst-Roussel)	Infections related to bone marrow transplantation	1991
Interleukin-2 (Proleukin)	Chiron/Cetus	Cancer	1992
Factor VIII (ReFacto)	Genetics Institute (Baxter)	Hemophilia	1992

* This was the first marketed drug based on monoclonal antibody technology.

Source: Adapted from H. Grabowski and J. Vernon, 'Innovation and structural change in pharmaceuticals and biotechnology', *Industrial and Corporate Change* 3, 2 (1994), based on 1991 survey report from Pharmaceutical Manufacturers Association and NDA Pipeline Reports 1983–1991.

and tissue plasminogen activator (t-PA), a treatment for blood clotting used mainly for stroke victims.

However, progress in launching new drugs was slower than had been predicted, prompting fears among investors that biotechnology had been over-hyped. Genetically engineered interferon, in particular, had been touted as a possible cure for cancer, and had figured prominently in the prospectuses issued by several biotechnology firms.[24] Although several types of interferon were approved for certain specific diseases, these were not the magic bullets that investors had been led to expect.[25] Several firms took up monoclonal antibody technology, but it proved difficult to develop antibody based drugs and only one—a treatment for organ transplant rejection developed by Ortho, a subsidiary of Johnson & Johnson—was approved during the 1980s.

Amgen, destined to become the largest and most profitable of the first generation firms, was founded in Thousand Islands, California, in 1980 and

---

[24] Rasmussen, *Gene Jockeys*, p. 129.

[25] Toine Peters, *Interferon, the Science and Selling of a Miracle Drug* (Abingdon: Routledge 2005).

floated 3 years later, just before the IPO window closed.[26] Its share price fell precipitately over the next 12 months, making it impossible to raise fresh capital from the public markets. Potential corporate partners were approached, but most pharmaceutical firms were sceptical about the value of biotechnology. Even Abbott, a pharmaceutical company which had supported Amgen's initial funding round, declined to be involved in its first drug programme, a treatment for anaemia which was then in late-stage trials. Abbott's chief scientist, like many of his counterparts in other pharmaceutical companies, was not enthusiastic about drugs based on large proteins which had to be injected rather than swallowed.[27]

Fortunately for Amgen, it was able to negotiate an alliance with a Japanese brewing company, Kirin, which paid $12m to acquire the Japanese rights to the anaemia drug.[28] Launched in the US in 1989 under the trade name Epogen, it soon became a blockbuster, a term that came to be applied to drugs which achieved annual sales in excess of $1bn. It was quickly followed by another big-selling drug, Neupogen (filgrastim), launched in 1991; Neupogen was used in cancer patients to reduce the risk of infection associated with chemotherapy.

Genentech had a shock in May 1987 when a panel of experts in the FDA declined to approve its blood clot dissolving drug, tissue plasminogen activator (t-PA), because of potential side effects. Although the drug was later approved, t-PA did not meet the company's market projections. Genentech's share price fell, as did those of several other biotech firms. These and other upsets unnerved investors. As one commentator wrote, 'Wall Street is sending a persistent message to its one-time favourites, the health biotech companies; you don't have the kind of future you thought you had'.[29]

Thus the progress of the US biotech sector in the 1980s was by no means smooth. It was a learning period not only for investors but also for managers, who had the difficult task of maintaining the creativity of their scientists while

---

[26] Thousand Islands was equidistant from the three universities from which Amgen recruited members of its scientific board: Cal Tech, UCLA, and UC-Santa Barbara. Walter W. Powell and Kurt Sandholtz, 'Chance, Nécessité et Naïveté, ingredients to create a new organisational form', in *The Emergence of Organisations and Markets*, edited by John F. Padgett and Walter W. Powell (Princeton: Princeton University Press 2012) p. 411.

[27] Gordon Binder and Philip Bashe, *Science Lessons, What the Business of Biotech Taught me about Management* (Boston: Harvard Business Press, 2008) p. 126. When Arthur Kornberg, a Nobel prize-winning biochemist, was trying to raise funds for a new company, DNAX, in 1981, he approached Roy Vagelos, president of Merck. 'Yet Vagelos revealed disdain for what DNAX and other small biotech ventures might achieve, and this disdain would dictate Merck policy for the next decade.' Arthur Kornberg, *The Golden Helix, inside Biotech Ventures* (Sausalito, CA: University Science Books, 1995) p. 57.

[28] Kirin was one of several Japanese non-pharmaceutical companies which were taking an interest in biotech (see Chapter 6).

[29] Gene Bylinsky, 'Bringing biotech down to earth', *Fortune* (7 November 1988).

at the same time steering them in commercially fruitful directions. The biotech chief executive had to be able to straddle both sides of the business.

The Management Challenge

Genentech was a rare case where the founder, Robert Swanson, who was not a scientist or a professional manager, took on the chief executive role and retained it for more than a decade. Swanson set clear priorities for the business, pressing his scientists to concentrate on marketable products while at the same ensuring that the firm generated enough revenue to cover its costs.[30] Other firms brought in experienced managers from outside. Bill Bowes, the venture capitalist who founded Amgen, had served on the board of Cetus and had seen the problems caused by that company's lack of clear commercial direction. While he assembled a distinguished scientific advisory board—a necessary step to impress investors—he appointed as chief executive a senior manager from Abbott, George Rathmann.[31]

By contrast, Chiron was run by the two founding scientists, William Rutter, head of the biochemistry department at UCSF, and Ed Penhoet, an associate professor at Berkeley. Rutter became chairman and Penhoet president, and they held these posts for several years, while retaining their university links. Rutter, who had run a large, inter-disciplinary university department, was able to bring together young scientists in a semi-academic environment—they were encouraged to publish in academic journals and to work closely with university researchers—while also focusing the business on products, principally vaccines, which had commercial potential.[32]

Biogen was dominated in its early years by a group of distinguished academics (including two from the UK) who had been brought together by venture capitalists to pool their expertise.[33] The company ran into financial problems in the mid-1980s, mainly because its research effort was spread over too many programmes. In 1985 the board appointed a new chief executive, James Vincent, who had been running Abbott's Diagnostics Division. Vincent cut out research programmes that had little commercial value, and put more weight behind the company's most promising molecule, beta interferon (Avonex), which was later approved as a treatment for multiple sclerosis.[34]

---

[30] Smith, *Genentech*, pp. 128–34.

[31] Powell and Sandholtz, 'Chance, Necessité et Naiveté', p. 411.

[32] Powell and Sandholtz, 'Chance, Necessité et Naiveté', p. 417.

[33] The two British scientists were Kenneth Murray from Edinburgh University and Brian Hartley from Imperial College.

[34] Lawrence M. Fisher, 'The rocky road from start-up to big-time player: Biogen's triumph against the odds', *Strategy and Business*, Issue 8 (1 July 1987).

Some of the most successful chief executives came from Baxter, a decentralized group in which divisional managers were given a high degree of autonomy to run their businesses in an entrepreneurial style.[35] One of the most successful of the 'Baxter boys' was Henri Termeer, who was brought in by its venture capitalist investors to run Genzyme two years after it was founded in Boston in 1981; Termeer, a Dutchman, had been head of Baxter's blood fractionating unit. He developed a business culture at Genzyme that was very different from that of Biogen. The emphasis was on prudent financial management, and on financing research out of internally generated funds rather than through partnerships. Genzyme later specialized in treatments for rare, so-called orphan diseases, where the patient population was small and where there was less likely to be competition from Big Pharma. Its first orphan drug was Ceredase, later replaced by Cerezyme (imiglucerase), an enzyme replacement therapy for the treatment of Gaucher's disease.[36]

Two other firms which turned to Baxter for their chief executives were Genetics Institute and Hybritech. The former, founded in Boston in 1980 to commercialize inventions made by two Harvard scientists, brought in Gabriel Schmergel, who had been head of Baxter's International Division. Schmergel took the company public in 1986, and remained chief executive for 16 years. Hybritech, founded by Ivor Royston and Howard Birndorf in San Diego, was one of the first to focus on monoclonal antibodies. Like Genentech, it was backed by Kleiner Perkins, and one of that firm's partners, Brooke Byers, was the company's first president and chief executive. Within a few months he persuaded Howard (Ted) Greene, who had been running Baxter's diagnostics division, to take his place.

Most of these first-generation firms, and the ones that followed in the late 1980s and early 1990s, came out of partnerships between academic scientists and venture capitalists; spin-offs from Big Pharma were rare. One Boston-based firm that broke the mould in this and other respects was Vertex, founded in 1989. The founder and driving force for the first fifteen years of its life was Josh Boger, who had been a senior scientist in Merck. Boger's decision to leave the company—he had become director of basic chemistry in his mid-thirties and was seen as a future head of research—came as a shock to the Merck hierarchy, but Boger believed that a small entrepreneurial firm would be a more effective vehicle for pursuing structure-based design, a technology which he believed would transform the drug discovery process.[37]

---

[35] Monica C. Higgins, *Career Imprints: How the 'Baxter Boys' Built the Biotech Industry* (San Francisco: Wiley, 2005).

[36] Gaucher's disease is a genetic disease caused by an enzyme deficiency. It can cause anaemia, low blood pressure, and enlargement of the liver and spleen.

[37] Barry Werth, *The Billion Dollar Molecule: One Company's Quest for the Perfect Drug* (New York: Simon & Schuster, 1994). The later history of Vertex is described in Barry Werth, *The Antidote: Inside the World of New Pharma* (New York: Simon & Schuster, 2014).

Boger's focus was not on recombinant DNA or monoclonal antibodies, and in that sense Vertex was a mini-pharmaceutical company rather than a biotechnology firm. But the definition of biotechnology, as noted in Chapter 1, came to be broadened to include any new drug discovery firm, irrespective of the technology that it used. Vertex, and other new firms that developed drugs based on synthetic organic chemistry, were classified as part of the biotech sector by industry associations, financial analysts, and investors.

The Emergence of Clusters

Several of the first generation firms developed proteins for therapeutic applications that were already known and understood, such as insulin for diabetes. Their novel contribution was the use of new manufacturing methods, made possible by recombinant DNA technology, that allowed much higher volumes of known molecules to be produced, tested, and, if shown to be safe and effective, brought to market. These first genetically engineered drugs came to be described as 'low-hanging fruit', although that phrase gives too little credit to the scientific and technical achievements of the pioneers.

These products enjoyed higher success rates in clinical trials than those of firms that came later. Many of the later entrants sought to identify molecules which would treat or cure diseases in novel ways. This generally involved a longer period of experimentation before the drug entered the clinic, and a greater risk of failure during clinical trials.

Monoclonal antibodies, in particular, proved difficult to convert into medicines. The first antibody drugs developed in the 1980s were derived from animal cells but patients' immune responses to these alien proteins made them more harmful than had been hoped. One of the pioneering firms, Centocor, suffered a near-catastrophic setback in 1992 when its lead drug failed to win approval from the FDA.[38] However, Centocor staged a remarkable recovery, launching two antibody-based drugs, ReoPro (abciximab) and Remicade (infliximab), in 1994 and 1999. By the end of the 1990s monoclonal antibodies were recognized as one of the fastest-growing segments of the biologics market.[39]

[38] Lara V. Marks, *The Lock and Key of Medicine: Monoclonal Antibodies and the Transformation of Healthcare* (New Haven: Yale University Press 2015) pp. 139–58.
[39] As noted in Chapter 1, 'biologics' is the term used to describe large-molecule drugs produced by biotechnology-based techniques.

As the biotech sector gained momentum, with another flurry of IPOs in the early 1990s, much of the innovative activity came to be concentrated in a few regional clusters, of which the most important were in San Francisco and Boston. These cities had two advantages in common: an established venture capital industry and an array of universities, research institutes, and teaching hospitals where scientists were working at the forefront of molecular biology. Both cities had benefited from the large-scale federal expenditure on military electronics during and after the Second World War; this had contributed to a wave of innovation in civil electronics, especially computers and semiconductors, much of it financed by venture capital.

The venture capital community grew faster in California than in Massachusetts, and the rate of new firm creation was higher, much of it in the form of spin-offs from existing companies. Fairchild Semiconductor, founded in 1956 by a team that broke away from an earlier semiconductor firm, was a prolific source of spin-offs.[40] This history of entrepreneurial dynamism made San Francisco a fertile environment for biotech firms.

Boston had a more conservative business culture than San Francisco, and there were also more reservations among academic scientists about participating in commercial activities.[41] Some universities, notably MIT, had long experience of working with industry, but in others there was unease over how closely an academic institution should be involved in a money-making enterprise. The spin-out of Genetics Institute from Harvard was preceded by a long internal debate, and the university decided not to take shares in the new company.[42]

By the 1990s these reservations had faded and scientists in Boston were as eager to be involved in biotech start-ups as their counterparts on the West Coast. Just across the Charles River from Boston, Kendall Square in the heart of Cambridge (Massachusetts) became a favoured location for biotech firms, venture capitalists, university technology transfer offices, and an array of service providers, all within a few minutes' walk of each other. Increasingly, too, the big pharmaceutical companies found it necessary to set up offices in Cambridge; they wanted to be close to the talent and the ferment of ideas coming out of an extraordinarily dynamic area. Even though some Cambridge biotech firms were later acquired by Big Pharma, the continuing flow of start-ups ensured that the entrepreneurial character of the cluster was preserved.

---

[40] Thirty-one semiconductor firms were founded in Silicon Valley in the 1960s, and most of them traced their lineage to Fairchild. Annalee Saxenian, *Regional Advantage* (Cambridge, MA: Harvard University Press 1994) p. 26.

[41] Walter W. Powell, Kelley Packalen, and Kjersten Whittington, 'Organisational and institutional genesis: the emergence of high-tech clusters in the United States', in Padgett and Powell (eds) *The Emergence of Organisations and Markets*, p. 444.

[42] Martin Kenney, *Biotechnology, the University-Industrial Complex* (New Haven: Yale University Press, 1986) pp. 78–83.

Of the other US regions, the one that came closest to matching Boston and San Francisco was San Diego.[43] The University of California at San Diego, like its sister university in San Francisco, had a strong biomedical science department, and the city was the home of two highly regarded research laboratories, the Scripps Research Institute and the Salk Institute. What it lacked in the early years was a local venture capital industry. The first major San Diego biotech firm, Hybritech, a specialist in monoclonal antibodies, was backed by Kleiner Perkins, and the others that were formed in the first half of the 1980s obtained finance from venture capital firms based in San Francisco or on the East Coast.

A boost for the San Diego cluster came from the takeover of Hybritech by Eli Lilly in 1985.[44] This was the first major Big Pharma acquisition of a small biotech; the price paid was about $300m, approximately 300 times Hybritech's 1984 earnings.[45] However, the Hybritech executives who joined Lilly after the merger found life in a large, Midwestern corporation not at all to their taste. Several of them were wealthy, thanks to the price that Lilly had paid for their Hybritech shares, and within a few months of the takeover they were looking for ways of resuming their former entrepreneurial careers. Ex-Hybritech managers founded or helped to found at least twelve companies formed in San Diego between 1986 and 1990.

Other regions tried to imitate San Francisco and Boston, but with only limited success.[46] US experience suggests that clusters emerge from the bottom up, through the spontaneous actions of different sets of players who work together for mutual benefit. State and local governments played no more than a permissive role in the growth of the San Francisco and Boston clusters. Top-down approaches, led by local government, had some success in a few areas, notably Research Triangle Park at Raleigh-Durham, North Carolina, but they lacked the intense interaction between universities, entrepreneurs, and financiers which characterized Boston and San Francisco.[47]

---

[43] Kjersten Bunker Whittington, Jason Owen-Smith, and Walter W. Powell, 'Networks, propinquity, and innovation in knowledge-intensive industries', *Administrative Science Quarterly* 54, 1 (March 2009) pp. 90–122.

[44] Steven Casper, 'How do technology clusters emerge and become sustainable? Social network formation and inter-firm mobility within the San Diego biotechnology cluster', *Research Policy* 36 (2007) pp. 438–55.

[45] Robert Teitelman, *Gene Dreams*, p. 179. Hybritech was using monoclonal antibody technology to make diagnostic kits but planned to develop therapeutics, a move that would require access to much larger amounts of capital.

[46] Joseph Cortright and Heike Mayer, *Signs of Life: The Growth of Biotechnology Clusters in the US* (Washington DC: Brookings Institution, June 2002). See also Scott Walsten, 'High-tech cluster bombs', *Nature* 438, 11 (March 2004).

[47] Lyman G. Zucker, Michael R. Darby, and Marylyn B. Brewer, 'Intellectual human capital and the birth of US biotechnology enterprises', *American Economic Review* 88, 1 (March 1998) pp. 290–306.

Support from Government

The two great biotech clusters were not created by government, either at the Federal or state level, and the same was true of the biotech sector as a whole. The principal protagonists were scientists, entrepreneurs, and financiers. Yet the government played a crucial supporting role.[48]

The most important contribution from government—for the life sciences industry as a whole, not just biotech—was the high level of public funding for biomedical research in universities, hospitals, and research institutes. The largest research funder has been the National Institutes of Health (NIH), which has more than tripled its budget in real terms since the 1970s. By the 2000s NIH spending on biomedical and healthcare related research had risen to around $30bn per year; it provides funds for potentially exploitable technologies, for the training of large numbers of scientists, and for sustaining an extensive research and development infrastructure.

The biotech sector has also benefited from investments by the Department of Defense (notably under the US Project Bioshield Act of 2004, which funded the development of vaccines and therapies to counter bioterrorism) and from investment in the Human Genome Project, which was started in 1990 with support from the Department of Energy. Not directly linked to biotech, but available for small, research-based firms across all sectors, was the Small Business Innovation Research Program and the Small Business Technology Transfer Program, through which government departments are required to allocate part of their research budget to small firms.[49] In the 1980s nearly half of the government's support for academic research went to the life sciences, significantly more than any other field. Annual US spending was several times that of countries such as Japan, Germany, and the UK.[50]

Government support for biomedical research was well established before the arrival of genetic engineering, but the early years of biotechnology saw several policy changes which contributed to the growth of biotech firms.

In 1979 the rules governing pension fund investment were changed in a way that stimulated the flow of funds into venture capital. Under the Employment Retirement Income Security Act of 1974 pension funds had been prohibited from investing substantial amounts of money in high-risk

---

[48] Some commentators attach greater importance to the role of government in US biotechnology: Steven P. Vallas, Daniel Lee Kleinman, and Dina Biscotti, 'Political structures and the making of US biotechnology', in *State of Innovation, the US Government's Role in Technology Development*, edited by Fred Block and Matthew R. Keller (Boulder: Paradigm Publishers, 2011).

[49] As an indication of the scale of this activity, since the year 2000 the NIH has awarded thousands of SBIR grants annually, each worth between $150,000 and $1,000,000. NIH STTR and SBIR grant distributions rose from around $130m in 1994 to over $600m in 2004. (http://grants.nih.gov/grants/funding/funding_program.htm).

[50] Cockburn et al., 'Pharmaceuticals and biotechnology'.

ventures. This was replaced in 1979 by a 'prudent man' rule which gave pension funds the freedom to invest in a wider range of assets. The result was a surge of funds into venture capital. In 1978, when $318m was invested in venture capital funds, individuals accounted for 32 per cent of the total. In 1988, when $3bn was committed to new funds, pension funds accounted for 46 per cent, by far the largest share, while the share of individuals had fallen to 8 per cent.[51] A growing proportion of this money flowed into funds that specialized in biotech.

The government also took steps to stimulate the commercialization of publicly funded research. This had been a contentious issue in earlier years, with some members of Congress arguing that intellectual property resulting from government-financed research should be placed in the public domain rather than monopolized through patenting. The opposing view was that commercialization would take place more quickly if universities were able to patent their inventions and license them to companies which had the capacity and the incentive to exploit them. Some patenting and licensing took place in the 1960s and 1970s, but these arrangements involved case-by-case negotiation between the universities and the funding agencies, and the rules governing patent rights were not uniform. Pressure from the universities and from industry led to the Bayh–Dole Act of 1980, which provided for a uniform federal patent policy and gave universities and small businesses rights to any patents resulting from government-funded research. The Act encouraged universities to set up technology transfer offices, following the example of MIT and others which had had close links with industry for many years.[52]

Of more direct relevance to the life sciences were two changes in the regulatory system. One was the introduction in 1983 of the Orphan Drug Act, designed to encourage firms to develop medicines for rare diseases; orphan diseases were defined as those that affected less than 200,000 people. For firms that developed orphan drugs, the Act provided a 7-year period of exclusivity, less burdensome clinical trial requirements, and tax incentives that partially offset the cost of research and development. Several biotech firms—Genzyme was an outstanding example—focused much of their development effort on orphan drugs.

The other regulatory change was the Hatch–Waxman Act of 1984.[53] This measure set out clearer rules on patent exclusivity and strengthened the

---

[51] Paul A. Gompers, 'The rise and fall of venture capital', *Business and Economic History* 23, 2 (Winter 1994).

[52] The change was widely seen as a step towards strengthening US industrial competitiveness at a time when competition from other countries, especially Japan, was intensifying, David C. Mowery, Richard R. Nelson, Bhaven N. Sampat. and Arvids A. Ziedonis, *Ivory Tower and Industrial Innovation, University–Industry Technology Transfer before and after the Bayh–Dole Act* (Stanford: Stanford Business Books, 2004) p. 92.

[53] The formal title of the Act was the Drug Price Competition and Patent Term Restoration Act.

ability of generic drug manufacturers to enter the market when patents expired. The effect on manufacturers of patented drugs was to create what came to be called the patent cliff, an immediate collapse in the price of the patented drug as soon as it lost patent protection; for companies that relied for the bulk of their profits on a single blockbuster drug, this could cause serious problems. However, many observers believe that the Act struck an appropriate balance between patent protection and encouraging generic entry. According to a recent assessment, 'The Hatch–Waxman Act was a significant policy success, simultaneously sharpening the incentives for breakthrough innovation, while ensuring diffusion and low-cost access after patent expiration.'[54]

Changing Relationships with Big Pharma

By the 1990s it was clear that the biotech sector was here to stay. What was less clear was how its relationship with Big Pharma would evolve. In the early days of 'biomania' there had been speculation that the new biotech firms might in due course displace the established pharmaceutical companies, just as entrepreneurial firms such as Intel had come to dominate the semiconductor industry at the expense of the older electronic component producers.[55] However, there were several reasons why a similar upheaval did not occur in pharmaceuticals.

First, most biotech firms were dependent on one or two drug candidates, and this made them more fragile than the well-financed and diversified pharmaceutical companies. If a Big Pharma drug failed in clinical trials, it might have little impact on the share price. For a biotech firm the financial consequences could be disastrous, particularly if it lacked revenues from products that were already on the market and was short of the capital needed to progress other drug candidates.

Second, although many biotech firms had promising early stage technologies they lacked the regulatory expertise and distribution networks which were crucial to appropriating the full value from their innovations.[56] To span the entire value chain from research to the point of sale—the FIPCO model (Fully Integrated Pharmaceutical Company)—called for skills and resources which were beyond the financial means of most biotech firms. Amgen, which achieved the unusual feat of producing and marketing two blockbuster drugs in quick succession, did it, but not many others could

[54] Cockburn and Stern, 'Finding the endless frontier'.
[55] Martin Kenney, 'Schumpeterian innovation and entrepreneurs in capitalism: a case study of the US biotechnology industry', *Research Policy* 15 (1986) pp. 21–31.
[56] Frank T. Rothermael, 'Incumbent's advantage through exploiting complementary assets via interfirm cooperation', *Strategic Management Journal* 22 (2001) pp. 687–99.

expect to do the same. Even Amgen had to share the revenue from its first products with commercial partners.

Third, by the end of the 1980s, following biotech firms' initial successes with a small but important set of novel products, the earlier scepticism in Big Pharma about biotechnology had given way to a recognition that the new techniques had opened up a promising approach to drug discovery. Some Big Pharma companies now began to consider how to absorb recombinant technology, and molecular genetic expertise more generally, into their own operations.[57] The simplest route was through licensing deals like the one Lilly had negotiated with Genentech. Another option was acquisition, but, as the Lilly/ Hybritech deal had shown, there was a risk that acquisitions by Big Pharma might lead to the exodus of the very people who constituted the principal assets of the acquired firm. As a senior Big Pharma executive remarked in the 1980s, 'they have a different type of culture. They don't want to feel they are part of a bureaucracy. They want a volleyball court. They want barbecues. They don't wear ties.'[58]

A different approach was a partnership that fell short of a complete takeover. In February 1990 Roche, the Swiss group, announced that it was buying 60 per cent of Genentech for just over $2bn, with an option to buy the remaining shares at a later date. The deal was driven on Genentech's side by a deteriorating financial situation caused in part by disappointing sales of t-PA. For Roche it was a way of tapping into a source of innovative science at a time when, following the expiry of patents on some of its best-selling drugs, it badly needed to rejuvenate its pipeline. Fritz Gerber, head of the Swiss company, was not a scientist (he had spent most of his career in Zurich Insurance), but he saw the importance of biotechnology and the need to participate directly in the stream of innovations that were coming out of the US. Aware of the special character of Genentech, he said he would change nothing in the way the Californian company was run; it retained its stock market listing and Roche took only two of the thirteen board seats.[59]

The Roche/Genentech deal was a sensational event, seen by one commentator as marking the biotech industry's loss of innocence.[60] Swanson had failed in his attempt to create a standalone, fully integrated pharmaceutical company, and it was reasonable to ask whether, if a firm as well-endowed as Genentech could not do it, others could expect to achieve that goal. On the other hand, the alliance was a model which other biotech firms might find

[57] Louis Galambos and Jeffrey L. Sturchio, 'Pharmaceutical firms and the transition to biotechnology: a study in strategic innovation', *Business History Review* 72, 2 (Summer 1998) pp. 250–78.

[58] Stuart Gannes, 'The big boys are joining the biotech party', *Fortune*, 6 July 1987.

[59] *Financial Times*, 5 February 1990.

[60] 'Ten deals that changed biotechnology', *Signals Magazine*, 17 November 1998.

attractive as they sought to strengthen their finances. Gary Pisano of the Harvard Business School, who was critical of the way biotech firms were financed, partly because of the pressure that was put on them by outside shareholders, commented that relationships like that between Roche and Genentech, if properly managed, could enable biotech companies to pursue longer-term R&D strategies, while benefiting from the intensive oversight of an informed investor.[61]

What followed after 1990 was a series of realignments as biotech firms that were under financial pressure sought to shore up their position. In some cases this involved mergers within the biotech sector. Cetus, the first of the dedicated biotechnology firms, had achieved some notable scientific successes, including the invention by Kary Mullis of polymerase chain reaction. This was a technique for generating a large number of copies of a particular DNA sequence; it became an essential tool in biological research and Mullis was awarded a Nobel Prize.[62] However, Cetus's record as a drug developer had been disappointing. By the end of the 1980s it had chosen to focus most of its efforts on developing a single drug, interleukin-2 (Proleukin), a treatment for cancer. When that drug was rejected by the FDA in 1990 Cetus was unable to raise further finance and it was bought by a US biotech competitor, Chiron, for $600m.

Another option was to do what Genentech had done with Roche. To the surprise of the industry, the Roche/Genentech deal did not lead to an exodus of Genentech scientists, and by the mid-1990s the company's drug pipeline was stronger than it had been ten years earlier; Herceptin, a treatment for breast cancer which became one of the most successful monoclonal antibody drugs, was launched in 1998. Another Swiss company, Ciba-Geigy (which later merged with Sandoz to form Novartis), made a similar arrangement with Chiron, whereby it acquired a 49.9 per cent stake for just over $2bn. William Rutter, Chiron's chairman, said the partnership would enable Chiron to do things that it could not do on its own: 'A small company like ours produces products that must and should be provided to the world's populations. The issue has been how to do it.'[63]

Three other first generation biotech firms, Centocor, Genetics Institute, and Immunex, all founded between 1979 and 1981, passed wholly or partly into the hands of pharmaceutical companies in the course of the 1990s.[64] From

---

[61] Gary P. Pisano, *Science Business* (Boston, Harvard Business School Press), p. 200.

[62] Paul Rabinow, *Making PCR: A Story of Biotechnology* (Chicago: University of Chicago Press, 2006).

[63] *New York Times*, 22 November 1994.

[64] Genetics Institute was acquired by American Home Products in 1996, Centocor by Johnson & Johnson in 1999. American Cyanamid bought a majority stake in Immunex in 1993; in the following year American Cyanamid was taken over by American Home Products, which sold its

that generation only Amgen, Biogen, and Genzyme were still fully independent at the end of the decade.

The Genomics Bubble and its Aftermath

An alternative strategy was to focus, not on drug discovery, but on creating broad technologies or platforms that could aid a range of client firms by providing tools to support their R&D efforts. Prominent in this group were the genomics companies, seeking to reveal the structure and function of genes and to use this knowledge in the search for promising molecular targets against which drugs could be designed. Other possible applications such as diagnostic tests were also commercially interesting. The potential importance of genomics in medicine was part of the rationale for the Human Genome Project, which was launched in 1990 by the National Institutes of Health and the Department of Energy, with support from governments and research institutions in several other countries, including the UK.

The genomics companies earned their revenue by selling information about genes. That Big Pharma might find such information valuable was highlighted by the agreement made in 1993 between Human Genome Sciences (HGS), one of the leading genomics firms, and the Anglo-American company, SmithKline Beecham.[65] William Haseltine, founder of HGS, described the deal as marking the transition from an industry based on chemistry to one based on genes.[66] The agreement, he said, 'put genomics on the map', opening an era in which knowledge about genes would be the key to product development in the life sciences. Another genomics firm, Millennium, signed a similar deal with Bayer whereby the German company, in return for an investment of $465m, would receive access to 225 new drug targets over a five-year period.

For a brief period at the end of the 1990s genomics firms were the darlings of the stock market. Investors believed that these firms held the key to opening up thousands of new avenues of drug research, and that belief contributed to an extraordinary stock market boom. This was in part an extension of the dot. com boom which had started a few years earlier, when investors fell in love with internet-related businesses and anything that seemed to be part of the 'new economy'. Biotechnology, and genomics in particular, was seen as 'the next big thing'. At the height of the boom in early 2000 the six leading

stake in Immunex to Amgen in 2002. American Home Products itself, which changed its name to Wyeth in 2002, was bought by Pfizer in 2009.

[65] SmithKline Beckman had merged with Beecham, one of the UK's four leading pharmaceutical companies, in 1989.

[66] Mark Edwards and Joan Hamilton, 'Ten deals that changed biotechnology', *Signals Magazine*, 17 November 1998.

US genomics firms were valued at $45bn.[67] Investors poured money not only into firms that specialized in genomics but also into others whose technology—for example, monoclonal antibodies—seemed on the brink of producing novel drugs.

Shares fell back in March 2000, when President Clinton and the British Prime Minister, Tony Blair, issued a joint statement indicating their hope that data arising from the Human Genome Project should be made freely available. This was taken to mean that genomics companies would no longer be able to patent their discoveries. At the same time there was a dawning realization among investors that the sequencing of the human genome was only the first step towards finding treatments for diseases that had a genetic origin. By the end of 2001 most of the firms that had gone public during the boom were trading at well below their flotation price. One by one, the leading genomics firms turned to developing their own drugs.[68] Several of them were later acquired by pharmaceutical companies.[69]

In 2000 no less than sixty-eight biotech firms went public, far more than in any previous boom. The collapse, when it came, was more painful, and had more lasting consequences. From the early 2000s the biotech sector found itself in a difficult financing environment which was to persist up to and beyond the world financial crisis of 2008–09. There was a revival in share prices in 2004–07, but the post-IPO performance of most of the firms that went public during those years was disappointing.[70] There were very few IPOs in 2008 and 2009, followed by the beginnings of a recovery in 2010.

In these circumstances, with very limited access to the stock market, companies which were already public had to decide whether they could raise enough finance to continue their drug programmes and remain independent, or whether they should look for a buyer. The period between 2000 and 2015 saw a series of mergers and acquisitions, some of them involving companies that had ranked among the sector's leaders (Table 2.4). In several of these cases, such as MedImmune and Genzyme, the target firm had suffered manufacturing or product missteps which had undermined the confidence of

---

[67] Paul Martin, Michael M. Hopkins, Paul Nightingale, and Alison Kraft, 'On a critical path: Genomics, the crisis of pharmaceutical productivity and the search for sustainability', in *The Handbook of Genetics and Society, Mapping the New Genomic Era*, edited by Paul Atkinson, Peter Glasner, and Margaret Lock (London: Routledge, 2009) pp. 145–62.

[68] H. Rothman and A. Kraft, 'Downstream and into deep biology: evolving business models in "top tier" genomics companies', *Journal of Commercial Biotechnology* 12, 2 (January 2006), pp. 86–98.

[69] Millennium was bought by Takeda in 2008, Human Genome Sciences by GlaxoSmithKline in 2012.

[70] Bruce L. Booth, 'Beyond the biotech IPO: a brave new world', *Nature Biotechnology* 27, 8 (August 2009), pp. 705–9.

Table 2.4. Mergers and acquisitions valued at over $5bn in US biotech, 2000–2015 (US companies except where otherwise stated)

Acquirer/target	Amount ($bn)	Year
Amgen/Immunex	10.0	2000
Biogen/Idec	7.5	2003
Novartis (Switzerland)/Chiron	5.1	2005
AstraZeneca(UK)/MedImmune	15.2	2007
Takeda (Japan)/Millennium	8.2	2007
Lilly/Imclone	6.5	2007
Roche (Switzerland)/Genentech	46.8	2009
Sanofi (France)/Genzyme	20.1	2011
Teva (Israel)/Cephalon	6.8	2011
Gilead/Pharmasset	11.2	2012
BMS/Amylin	5.3	2012
Amgen/Onyx	10.4	2013
Merck/Cubist	9.5	2014
Shire (Ireland)/NPS Pharmaceuticals	5.2	2015
Abbvie/Pharmacyclics	21.0	2015

Source: HBM Pharma/Biotech M & A Report 2014, HBM Partners AG, Zug, Switzerland.

investors and made them vulnerable to takeover.[71] The acquirers included non-American companies such as AstraZeneca from the UK and Takeda in Japan, which were seeking to strengthen their position in biologics and to enlarge their R&D footprint in the US.[72]

Although finance for start-up and early-stage firms was harder to find than it had been in the 1990s, there were still scientists and entrepreneurs who had ambitions to build another Amgen or another Genentech, and investors willing to support them. In 2002 two Boston-based venture capital firms, Cardinal Partners and Polaris, launched a new biotech company, Alnylam, to work in the promising field of RNA Interference.[73] The technology was licensed from a group of institutions, including MIT in the US and the Max Planck Institute in Germany, and the investors recruited John Maraganore, who had held senior posts in Millennium and Biogen, to be chief executive. As Maraganore remarked later, 'this was a negative period for early stage venture investment: most VCs were focusing on later stage assets often acquired from pharmaceutical companies. But the discovery of RNAi was so potentially

---

[71] In the case of MedImmune, disappointing sales of the company's nasal flu vaccine, FluMist, caused the share price to fall, prompting the activist investor, Carl Icahn, to buy a stake in the company at the end of 2006; a few months later MedImmune put itself up for sale.

[72] AstraZeneca had been formed in 1999 by a merger between Zeneca (the pharmaceutical business which had been demerged from Imperial Chemical Industries in 1993) and Astra of Sweden. AstraZeneca's first big move into biologics had come in 2006 when it acquired a leading British biotech firm, Cambridge Antibody Technology (discussed in Chapter 3).

[73] RNA interference is a process in which the introduction of double-stranded DNA into a cell inhibits the expression of genes.

disruptive, many scientists believed it could be hugely important.'[74] Alnylam went public in 2004, and raised further sums through partnership deals with Novartis, Roche, and Takeda.

Alnylam relied on the classic financing model—support from venture capital followed by further stock market funding—and there were a few others which followed the same route. However, in 2008 and 2009, when the financial crisis was at its height, optimism was in short supply. As a Big Pharma executive commented in 2009, 'There are now more than 300 public biotech firms and hundreds more private firms. And I would guess that maybe half of them have 6 months before they run out of cash. These firms are firing people and stopping projects that may have merit in an effort to preserve cash. And with the drying up of traditional sources of public and private capital, the only solution, short of sale to another biotech firm or Big Pharma, is venture capital. But because of investors' reduced appetite for risk, the VCs themselves are now strapped for funding.'[75]

The retreat of the venture capitalists was partly offset by Big Pharma's venture capital arms, such as Johnson & Johnson Development Corporation and GlaxoSmithKline's SR One, which were still willing to invest in early-stage ventures.[76] Unlike traditional VCs, they were not required to exit their investments within a specified period in order to return funds to their investors. Nevertheless, for biotech firms that wanted to grow, access via the stock market to the vast pool of capital offered by public investors was essential.

The Revival of Investor Support

In 2010 and 2011, as the US economy began to recover, a few biotech companies floated on NASDAQ, while others were sold for high prices to pharmaceutical companies. That Big Pharma still had an appetite for acquisitions was comforting for investors, but there was one acquisition which aroused some concern in the biotech community. This was the takeover of Genzyme by Sanofi, the French company which had become one of the world's largest pharmaceutical groups through the takeover of Aventis in 2004.[77]

---

[74] Interview with John Maraganore, 11 June 2014.

[75] Judy Lewent, chief financial officer of Merck, quoted in 'Life sciences roundtable: strategy and finance', *Journal of Applied Corporate Finance* 21, 2 (Spring 2009) pp. 8–35.

[76] SR One was founded by SmithKline Beckman in 1985, and became part of GSK after the merger between GlaxoWellcome and SmithKline Beecham in 2000.

[77] Aventis had been formed in 1998 by the merger of the life science businesses of the German company, Hoechst, and Rhône-Poulenc of France. The other big French pharmaceutical company, Sanofi, launched a bid for Aventis in 2004; there was a rival offer from Novartis, but opposition from the French government caused the Swiss company to withdraw.

Genzyme had lost investor support in 2009 as a result of manufacturing problems in one of its factories. The falling share price prompted the activist investor, Carl Icahn, to buy a stake in the company and to press for changes in strategy and for the removal of the long-serving chief executive, Henri Termeer. In 2011, after an acrimonious battle, Icahn achieved part of what he wanted when Sanofi took over the company for $20bn.

While this was a good outcome for investors, there was anxiety in Boston that the absorption of Genzyme into Big Pharma would dilute its innovative culture and drain some of the dynamism from the biotech cluster which it had helped to create.[78] 'The loss of Genzyme does matter', one venture capitalist remarked. 'It was one of the few companies with credible local leadership, credible at the national level. Local leaders are important—people like Termeer at Genzyme and Schmergel at Genetics Institute have been part of the community and advocates for the sector. Someone like Mark Fishman (head of the Novartis Institutes for Biomedical Research, set up in Cambridge in 2002) cannot play the same role.'[79]

The other big Boston biotech, Biogen Idec (Biogen had merged with Idec in 2002), also came near to losing its independence. In 2008 and 2009 activist investors led by Carl Icahn criticized the company for its lacklustre record on new products and pressed the management either to sell the business or to split itself into two, an oncology unit and a neurology unit.[80] The expectation in Boston was that Biogen Idec would have to find a pharmaceutical acquirer. However, in 2010 the company appointed a new chief executive, George Scangos, who had previously run a smaller biotechnology firm, Exelixis. The change of management, together with the imminent approval by the FDA of several potentially big-selling drugs, appeared to remove the takeover threat.[81]

The takeover of Genzyme and the near-takeover of Biogen Idec showed that even the largest biotechs were not immune to shareholder attack if their performance deteriorated. But these events did not imply any diminution in the dynamism of the US biotech sector, or in the ambition of younger firms to build large, free-standing businesses.

One of the strengths of the US biotech sector is the number of mid-tier firms from which new top-tier players can rapidly emerge. The most spectacular rise in recent years has been that of California-based Gilead. Founded in 1987, floated in 1992, and led from 1996 by John Martin, Gilead based its early growth on antiviral medicines, including Viread (tenofovir), a treatment for HIV. From 1999 Martin embarked on a series of acquisitions which took the

[78] *Nature* 470, 449 (2011).
[79] Interview with Marc E. Goldberg, BioVentures Investors, 9 June 2014.
[80] *BioCentury*, 11 January 2010.
[81] Biogen's Tecfidera (dimethyl fumarate), an oral drug for the treatment of multiple sclerosis, was launched in 2013 and quickly achieved blockbuster status.

company into treatments for cardiovascular disease and respiratory disease. Its most daring deal came in 2011 when it spent $10bn to buy Pharmasset, a biotech firm which was developing an oral drug for hepatitis C, designed to cure rather than treat the disease. Launched in 2013, the new drug, branded as Sovaldi (sofosbuvir), was a huge success.

The price which Gilead charged for Sovaldi—$94,500 for a twelve-week course of treatment—prompted protests from politicians and doctors, raising fears in the industry that America's free-market pricing model might be in jeopardy.[82] Nevertheless, what Gilead showed was that the creation of another Amgen through a combination of acquisitions and in-house research was still possible: the success of Sovaldi lifted Gilead's market capitalization above that of some long-established Big Pharma companies.

Gilead's performance was exceptional, but its success coincided with a broader revival of investor interest in biotech, beginning in 2013 and accelerating over the next 2 years. The recovery in share prices was driven in part by macro-economic factors; low interest rates reduced the cost of capital for biotech firms, and attracted investors to an industry which offered the prospect of a high return. But there were other reasons that justified a re-rating of biotech firms.

In a study published in October 2012 under the title 'glorious middle age', a leading US analyst, Geoff Porges of Bernstein Research, noted, first, that the public biotech sector had finally achieved sustained profitability after more than 30 years of losses (Figure 2.2), and, second, that investors could look

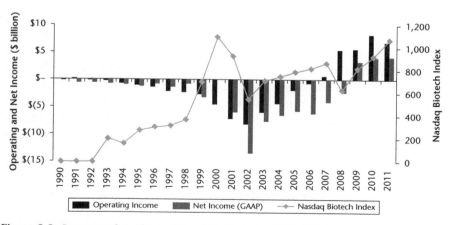

Figure 2.2. Losses and profits in US public biotech, 1990–2001
Source: Bernstein Research.

[82] *Financial Times*, 11 February 2015.

forward to a further period of improved performance.[83] This was partly due, Porges argued, to greater financial discipline among the leading companies; they were paying more attention to cost control, margin improvement, and the efficient allocation of capital. But there were other factors which justified an optimistic view: a more favourable regulatory climate, with an increase in the number of FDA approvals, especially for orphan drugs;[84] the development of speciality drugs for severe diseases, including targeted cancer therapies and treatments for hepatitis C; the likelihood that more of these drugs would become 'mega-blockbusters', with annual sales exceeding $2bn; and the prospect of increasing sales in emerging markets.[85]

A contributory factor was the recognition by investors that some technologies which had been under development for several decades were showing renewed commercial promise. One of several firms backed by Third Rock Ventures, a Boston VC founded by Mark Levin, formerly chief executive of Millennium, was Genetix, a specialist in gene therapy. Genetix had been founded in 1992 and had struggled to stay afloat during a period when gene therapy was out of favour with investors. By 2010 it had advanced to the point where, in Third Rock's view, it offered the prospect of a major medical advance and justified new investment. The company was refinanced by a syndicate led by Third Rock, the name was changed to bluebird.bio, and a Third Rock executive, Nick Leschly, was installed as chief executive. Other venture capital firms joined the syndicate later, and, taking advantage of the revival of investor interest in gene therapy, the company also attracted major institutional investors such as Fidelity and Deerfield. Bluebird.bio raised $101m at its IPO in June 2013.

During 2013 forty-one biotech firms were floated, raising a total of $3.5bn, the highest one-year total since 2000, and the surge continued into 2014, when seventy-four firms were listed, raising $5.3bn.[86] The NASDAQ biotechnology index rose by 66 per cent in 2014, compared to a 26 per cent rise for the Dow Jones Industrial Average. A striking feature of the 2014 cohort of biotech IPOs was the number of newly listed firms that were at an early stage in their drug development programmes. Nine firms whose drug candidates were in the preclinical stage were floated in 2014 compared to only one in 2013; twenty firms which were in Phase I were listed, compared to eight in 2013. Investors had recovered their appetite for high-risk biotech ventures.

[83] 'Glorious middle age: the 2012 biotech rally, why it might continue and how to participate', *Bernstein Research*, October 2012.

[84] The FDA approved 41 new therapeutics in 2014, including 11 biologics. Approvals were up by 50 per cent from the 27 approved in 2013, and up 30 per cent from the 5-year average of 31.6 per year. *Nature Reviews Drug Discovery* 14 (2015) pp. 77–81.

[85] The picture painted by this report is based on an overview of several hundred public firms, and while the sector as a whole had reached sustained profitability, this was based on the revenues of a minority. Most biotech firms continued to make losses.

[86] Ernst & Young, *Beyond Borders*, 2014.

Table 2.5. Top US biotech companies, measured by market capitalization on 13 July 2015

Company	Date of foundation	Date of IPO	Market capitalization ($bn)
Gilead	1987	1992	167.2
Amgen	1980	1983	117.2
Celgene	1986	1987	94.3
Biogen*	1978	1983	91.3
Regeneron	1988	1991	54.3
Alexion	1992	1996	43.5
Vertex	1989	1991	30.0

Source: Google Finance.
Biogen, formerly Biogen Idec, dropped Idec from its name in March, 2015.

The US Biotech Sector in 2015

As Porges had pointed out in his 2012 study, only 20 per cent of all US public biotech companies were profitable, and most of them were exposed to the risk of failures in clinical trials. In that sense the industry retained some of the characteristics that had been apparent in the 1980s and 1990s. But with the emergence of four large, profitable companies at the top end (Amgen, Biogen, Celgene, and Gilead), the structure of the industry was less fragile than it had been.[87] Below the big four there was a group of sizeable companies, including Regeneron, Alexion, and Vertex (Table 2.5), and below them more than a score of firms valued in mid-2015 at between $10bn and $30bn.

It had taken Gilead 15 years and several changes in strategy to go from start-up to profitability, and another 10 years to achieve 'big biotech' status. Given the high costs involved, the appetite of Big Pharma for firms with attractive assets, and the fact that the vast majority of drugs fail to reach the market, it would be hard for younger companies to retain their independence over so long a period. Nevertheless, there were some which had aspirations to do so. Alnylam, for example, a developer of RNAi therapeutics, continued to do big deals with partners (including a $700m deal with Sanofi/Genzyme at the start of 2014), while progressing its own drugs through clinical trials. By the end of 2014 it had raised some $2bn since foundation, of which $1.2bn had come from

---

[87] Celgene had entered the biotechnology sector by an unusual route. Originally a subsidiary of Celanese Corporation, it had been spun out as a separate company in 1986 and floated on the stock market. In 1987 it later acquired the worldwide rights to thalidomide. This drug had a toxic reputation dating back to the 1950s and 1960s; used as a treatment for morning sickness in pregnant women, it had produced serious birth defects. However, researchers at Rockefeller University had shown that thalidomide could treat the skin lesions caused by leprosy and might also be applied to other life-threatening diseases. In 1998 Celgene secured FDA approval to market thalidomide, under the trade name Thalomid, for treating leprosy, and this was followed within a few years by Revlimid (lenalidomide), which was targeted at cancer. Like Gilead, Celgene built on these early successes by making a series of acquisitions, mainly in oncology.

partnership deals and the rest from issuing new equity. In mid-2015 Alnylam, which was still loss-making, was valued in the stock market at just over $10bn.

Alnylam was one of several firms which had a Big Pharma partner as a significant shareholder—in this case Sanofi with just over 12 per cent. There were some similarities between these arrangements and the Roche/Genentech deal in 1990, except that Big Pharma had only a minority stake and in most cases there were provisions that prevented the partner from making a bid for the whole company. As a way of organizing collaboration between Big Pharma and small biotech, this 'dominant minority ownership' model was less disruptive for the management of the acquired firm than an acquisition. With or without such alliances, some of the newly floated biotechs had the ambition and the financial capacity to grow into standalone, product-based companies.

The surge in biotech IPOs had begun after a long period when capital for biotech firms was in short supply, and this had led to the disappearance of weaker firms. Those that survived the famine were of higher quality and likely to attract investors as soon as sentiment in the stock market turned. Much of the money that went into biotech IPOs between 2012 and 2015 was directed towards a limited set of treatment areas—the largest recipients were oncology companies and companies targeting orphan diseases—where the prospects for breakthrough innovations, based on sound scientific evidence, looked good.

Some of the valuations seemed excessive, as did the prices paid in some M&A transactions.[88] Setbacks among the high-flying firms could damage investor confidence in the sector, as had happened in the past. There were other clouds on the horizon, not least the growing public concern about the industry's pricing practices—an issue taken up by Hillary Clinton as part of her campaign for the Democratic Party's presidential nomination. Worries about government intervention in pricing, together with fears of a possible rise in interest rates, contributed to a sell-off of biotech stocks in the autumn of 2015.

Whether this was a temporary setback or the beginning of an end to the long-running NASDAQ biotech boom was a matter of much debate when this book went to press. What was clear, nevertheless, was that the strength of investor support in the 2013–2015 period had reinforced the position of the US as the global centre of biotech innovation and investment. The European biotech sector did share to some extent in the revival of investor interest; there were fifteen biotech IPOs in Europe in 2014, raising a total of €719.5m, compared with only five in 2013, raising €59.4m.[89] Yet even the 2014 figure

---

[88] An extreme example of Big Pharma's eagerness to acquire promising biotech firms was the bidding contest for Pharmacylics, developer of Imbruvica (ibrutinib), a recently approved drug promising blockbuster sales from significantly improving the treatment of certain rare blood cancers, in March 2015; the winner was AbbVie, which paid $21bn.

[89] In 2014 seven European firms chose NASDAQ rather than their home exchanges for their IPO: two from The Netherlands (UniQure and ProQR Therapeutics), two from Germany (Affimed and

Table 2.6. European and US public biotech sectors in 2014

	Europe	US
No of public companies	150	402
Market capitalization	€66.0bn	€633bn
Turnover (2013)	€10.2bn	€71.9bn
Capital raised via IPOs (2009–2014)	€1.02bn	€8.3bn
No of IPOs (2009–2014)	41	141

Source: Biocom AG, Comparative analysis of European biotech stock markets, 2015 Facts and Trends.

was meagre compared to the US. Between 2009 and 2014 the amount of capital raised through IPOs was eight times higher in the US than in Europe (Table 2.6).

American dominance could be measured, not only by IPOs and the number and size of US biotech firms, but also by the contribution these firms had made to the supply of innovative drugs. According to an analysis of the period between 1998 and 2007, biotech firms were responsible for the initial development of most of the university-discovered drugs that were scientifically novel or offered substantial benefit over existing drugs. Moreover, the discovery of new therapeutic biologics throughout the world during that period was dominated by US firms and US universities that partnered with them. The study concluded that without the contribution of mainly US-based biotech firms the number of innovative drugs that respond to unmet medical needs would have been substantially lower.[90]

Sources of US Strength in Biotech

That US firms were the first-movers in biotechnology might be regarded, at least in part, as a matter of luck—the fact that recombinant DNA was invented in the US and quickly identified by American scientists and entrepreneurs as a superior process for making proteins such as insulin. More relevant, for countries trying to learn from the US experience, is what happened next—how the early start was built on and extended, establishing a momentum of growth which withstood occasional setbacks and periodic famines in the supply of capital.

Pieris), one from Switzerland (Auris), one from Denmark (Forward Pharma), and one from Ireland (Innocoll).

[90] Robert Kneller, 'The importance of new companies for drug discovery: origins of a decade of new drugs', *Nature Reviews Drug Discovery* (November 2010).

Key elements in the US biotech ecosystem were: the quality and quantity of academic research in biomedical science, much of it funded by the Federal government; effective arrangements for transferring academic discoveries with commercial potential into profit-making enterprises; and a financial system which was well equipped to support growing firms in science-based industries. The existence of an established pharmaceutical industry provided a readily available and substantial source of managerial talent and commercial partnerships.

A fifth key element was that American firms also had the advantage of operating in a large and receptive market.[91] Small countries can do well in science-based industries (as Switzerland has done in pharmaceuticals), but the size and character of the US domestic market is relevant to an understanding of US success in biotech. Although the regulatory arrangements for drug safety and efficacy are broadly similar in the US and Europe, the rate of adoption of novel drugs is quicker in the US. Once a drug has been approved by the European Medicines Agency it has to go through country-by-country pricing and reimbursement procedures which may take many months. In the US, drugs are launched as soon as they have been approved by the FDA, and the price is set by the manufacturer or negotiated with health insurers; the government, which buys drugs for the Medicare and Medicaid programmes, generally follows the commercial price. These factors make the US a more receptive and lucrative market for innovative drugs than Western Europe or Japan.

Whether the amount of money spent on drugs in the US is too large, reflecting the lobbying power of the pharmaceutical industry, has long been a matter of controversy and there were signs in 2015, as protests over high drug prices intensified, that the industry might be losing political support. It is not clear whether the long-standing hostility in the US to European-style 'socialized medicine' will be strong enough to deter the government from imposing some limits on the pricing freedom enjoyed by the drug manufacturers. There has also been criticism of the role played by pharmaceutical firms in boosting the demand for medicines by the introduction of so-called life style drugs which have little therapeutic value.[92] This is one of several aspects of a healthcare system which is widely regarded as wasteful; healthcare spending in the US as a percentage of GDP is much higher than in other countries, without achieving significantly better outcomes.[93]

---

[91] The importance of these five elements is discussed in prior studies: Martha Prevezer, 'Ingredients in the early development of the U.S. Biotechnology Industry', *Small Business Economics* 17 (2001) pp. 17–29; Rebecca Henderson, Luigi Orsenigo, and Gary Pisano, 'The pharmaceutical industry and the revolution in molecular biology: interactions among scientific, institutional, and organisational change', in *Sources of Industrial Leadership: Studies of Seven Industries*, edited by David C. Mowery and Richard R. Nelson (Cambridge: CUP, 1999): pp. 267–311.

[92] John Abraham, 'Pharmaceuticalization of society in context', *Sociology* 44, 4 (2010) pp. 603–22.

[93] In 2013, according to World Health Organisation figures, spending on health as a proportion of GDP was 17.1% in the US, 11.7% in France, 11.3% in Germany, 10.3% in Japan, and 9.1% in

For other countries, learning from the US in biotech did not require adopting an American approach to the provision of healthcare. What mattered to governments seeking to generate employment and export revenues was whether they could create, within their particular political and social systems, an environment which encouraged the growth of innovative drug discovery firms. These efforts are the subject of the next four chapters.

the UK. For a comparison between the American and other healthcare systems see Karen Davis, Kristof Stremikis, David Squires, and Cathy Schoen, 'Mirror, mirror on the wall: how the performance of the US health care system compares internationally' (New York: The Commonwealth Fund, June 2014).

3

The British Response

Our scientists have played a prominent and at times dominant role in the establishment of modern biology, particularly molecular biology; British industry must not allow the industrial advantage of that position to be frittered away.
Spinks Report 1980[1]

The rise of Genentech and the other biotechnology pioneers in the US coincided with a period of anxiety in the UK about the country's loss of ground in high-technology industries. These were the industries on which the UK would have to rely as older industries declined, and in most of them there were few if any British-owned firms capable of competing against the world leaders. The Labour governments which held office between 1964 and 1970 and again between 1974 and 1979 tried to strengthen the UK's position in advanced technology by building larger companies. In computers, for example, International Computers Limited (ICL) was formed in 1968 by a series of government-induced mergers in the hope that it would become a credible rival to IBM. However, towards the end of its second term Labour came to recognize that one of the keys to American success in high technology lay in an environment which encouraged the creation of ambitious, entrepreneurial firms, more nimble than large, established corporations.

In 1978 the National Enterprise Board (NEB), a government agency whose remit was to promote the modernization of British industry, sponsored the creation of Inmos, a semiconductor firm whose technology was thought to be on a par with that of American manufacturers. The hope was that Inmos would establish itself in the mainstream of the semiconductor industry, alongside US companies such as Fairchild and Intel.[2]

---

[1] *Biotechnology*, report of a joint working party of the Advisory Council for Applied Research and Development, Advisory Board for the Research Councils, the Royal Society, generally referred to as the Spinks Report (London: HMSO, 1980), para 3.2.
[2] W. B. Willott, 'The NEB involvement in electronics and information technology', in *Industrial Policy and Innovation*, edited by Charles Carter (London: Heinemann, 1981).

The government saw semiconductors as an enabling technology which would have wide application, not just in electronics, but in many other industries. By the time Inmos was created, biotechnology was being viewed in the same light. The belief that in this field the UK could be a world leader was reinforced by the contribution which British universities and research institutes had made to the underlying science of molecular biology. It was in Cambridge, at the research laboratory set up by the Medical Research Council, where James Watson and Francis Crick had worked out the structure of DNA in 1953. They were given Nobel prizes, together with Maurice Wilkins from King's College London, in 1962. In the same year two other Cambridge scientists, John Kendrew and Max Perutz, shared the Nobel Prize in chemistry for their work on the structure of proteins. It was in the same laboratory, relocated to the outskirts of Cambridge and renamed the Laboratory of Molecular Biology, where Georges Köhler and César Milstein did their work on monoclonal antibodies; they too won Nobel prizes, as did Fred Sanger for his work on DNA sequencing.[3]

The events that followed the Köhler–Milstein discovery came to be seen as an example of the UK's chronic failure—especially in comparison with the US—to turn inventions into commercially valuable products.[4] The National Research Development Corporation (NRDC), the government agency which at that time was responsible for commercializing inventions resulting from publicly-funded research, considered the case for patenting monoclonal antibody technology, but did not do so on the grounds that it could not identify any immediate practical applications.

This decision was later criticized by Milstein and other scientists. Comparisons were drawn with the 'loss' of penicillin during the Second World War. Discovered by Alexander Fleming in 1929 and later converted into a usable medicine by scientists at Oxford University, penicillin was made available to the US as a war-time measure and subsequently exploited mainly by US pharmaceutical companies. The penicillin episode was partly responsible for the creation of the NRDC in 1948; its aim, as one newspaper put it, was to 'stop foreigners filching our ideas'.[5] However, in the case of monoclonal antibodies, the technology was used first as a research tool and in diagnostics,

---

[3] The MRC's Unit for the Study of Molecular Structure of Biological Systems, created in 1947, was housed in the Cavendish Laboratory until 1962, when the new Laboratory of Molecular Biology was established. Soraya de Chadarevian, *Designs for Life, Molecular Biology after World War II* (Cambridge: CUP, 2002) pp. 228–32.

[4] The obvious contrast is with the Cohen–Boyer Recombinant DNA patents, licensed widely to generate over $250 million in revenues for Stanford University and the University of California, San Francisco. Doogab Yi, *The Recombinant University: Genetic Engineering and the Emergence of Stanford Biotechnology* (Chicago: Chicago University Press, 2015).

[5] *Daily Herald*, 14 April 1948, quoted in Robert Bud, *Penicillin, Triumph and Tragedy* (Oxford: OUP, 2007) p. 72.

and it took many years of further development before big-selling drugs based on monoclonal antibodies were launched on the market. Given the slow and difficult development of monoclonal antibody drugs, it is doubtful whether the patenting of the Milstein–Köhler discovery by the NRDC would have given UK-based drug discovery firms a competitive advantage.

Meanwhile a group of British scientists who were aware of the potential importance of biotechnology was trying to generate support in the science community and in Whitehall for a government initiative in this field.[6] The group included Sydney Brenner at the Laboratory of Molecular Biology (LMB), a future Nobel Prize winner, and Brian Hartley at Imperial College, a co-founder of Biogen.[7] The outcome was the establishment at the end of 1978 of a committee, made up of scientists, industrialists, and government officials, to examine the potential of the new techniques. The chairman of the committee was Alfred Spinks, head of research at Imperial Chemical Industries (ICI).

The Creation of Celltech

The Spinks report, published in 1980 not long after Margaret Thatcher's Conservative government had come to power, called for urgent action to ensure that the UK did not lose ground to other countries in biotechnology.[8] A concerted approach was needed from government and industry, and 'a commitment, financial and otherwise, at least comparable to that taken or contemplated by our major international competitors'.[9]

The committee recommended greater coordination among the various bodies concerned with scientific research—the Research Councils (through which the government channelled funds to academic laboratories), the NRDC, and the Department of Industry —to provide 'a firm commitment to strategic applied research' in biotechnology.

The committee was pessimistic about the ability of the private sector on its own to provide the kick-start that was necessary. British firms seemed less able or less willing to take advantage of the opportunities in biotechnology than their counterparts in other countries; they also had less access to venture capital on the scale that was available in the US. In a proposal that was

---

[6] Robert Bud, 'From applied microbiology to biotechnology: science, medicine and industrial renewal', *Notes & Records of the Royal Society*, 64 (July 2010) pp. 17–29.

[7] Sydney Brenner, one of the UK's leading molecular biologists, was director of the LMB from 1979 to 1986. He shared the Nobel Prize for Physiology or Medicine in 2002.

[8] *Biotechnology*: Report of a Joint Working Party of the Advisory Council for Applied Research and Development, the Advisory Board for the Research Councils, the Royal Society (London: HMSO, 1980).

[9] Spinks Report Para 4.5.

bound to be politically controversial, the committee suggested that the NEB and the NRDC should consider the establishment, with public funds, of a new research-oriented biotechnology firm of the kind that was taking shape in the US.[10]

The Prime Minister was strongly opposed to state intervention in industry and eager to dismantle the National Enterprise Board—Inmos, the NEB-backed semiconductor firm, was sold in 1984—but in the case of biotechnology she agreed that a new company could be set up, as long as the private sector was fully involved; Margaret Thatcher, who had a degree in chemistry from Oxford University and first-hand experience of industrial R&D, saw the initiative as a way of encouraging closer collaboration between academic laboratories and industry.

Celltech was created at the end of 1980, with 44 per cent of the equity held by the NEB and the rest by four private-sector institutions: Prudential Assurance, one of the UK's biggest institutional investors; Midland Bank; British and Commonwealth Holdings, a financial services group whose ambitious chief executive, John Gunn, was excited by the potential of biotechnology; and Technical Development Capital, which was part of the Industrial and Commercial Finance Corporation (ICFC). ICFC had been set up by the clearing banks after the war at the request of the Bank of England to improve the flow of finance for small and medium-sized companies. Through Technical Development Capital, which it acquired in 1966, ICFC took on an additional role as a provider of equity to technology-based enterprises.[11]

The technical basis for Celltech was to be privileged access to the LMB's research in recombinant DNA and monoclonal antibodies. This was a departure from the previous system, whereby all discoveries arising from Medical Research Council (MRC)-funded research were passed to the NRDC. Sydney Brenner had been actively seeking a means of commercializing the laboratory's research and was a strong supporter of the Celltech plan.[12]

The first managing director was Gerard Fairtlough, a Cambridge-trained biochemist who had been head of Shell Chemicals UK before joining the NEB as an adviser; most of his previous career had been in general management.[13] One of his first appointments, as head of research and development, was Norman Carey. Carey had been working at the British subsidiary of G. D. Searle, an American pharmaceutical company, and was experienced in

---

[10] Spinks Report Para 4.14.

[11] Richard Coopey and Donald Clark, *3i, Fifty Years Investing in Industry* (Oxford: OUP, 1995) pp. 84–7.

[12] Professor William Shaw, a member of Celltech's Science Council, described the decision to set up the new company as 'a breathless response to the failure to patent monoclonal antibodies'; Mark Dodgson, *Celltech, The First Ten Years of a Biotechnology Company* (Brighton: University of Sussex, 1990), p. 89, note 15.

[13] The early history of Celltech is covered in Dodgson, *Celltech*.

the use of recombinant DNA techniques. Searle's UK laboratory, set up in 1966, had become a leading centre of research in molecular biology; Carey had been offered a job in Genentech but had turned it down. Celltech established its headquarters and research laboratory in Slough (not in Cambridge as had been the original intention), conveniently close to the Searle laboratory in High Wycombe. Several other Searle scientists joined Carey soon after Celltech was set up.[14]

Celltech also recruited managers from outside the pharmaceutical industry who had commercial experience. As one of them, John Berriman, recalled, 'I wanted to work in a growth business, and at that time there were only two growth industries: information technology and biotech'.[15] Berriman, who had been working in the engineering industry, joined in 1982, and was later given responsibility for what was to become a major source of revenue for Celltech, the manufacture of monoclonal antibodies for supply to other companies.

Celltech was the UK's first biotechnology firm and by far the largest, in terms of scientists employed. It came to be regarded as the flagship of the industry, until it was acquired by UCB, a Belgian company, in 2004. Several of its early recruits went on to hold senior positions in other biotech firms, and in venture capital. Berriman, after leaving Celltech, worked as a director of Abingworth, one of the leading British venture capital firms, and served as chairman or director of several biotech start-ups. Andrew Sandham, who was head of business development at Celltech from 1984 to 1989, was involved in no less than twelve start-ups, in the US as well as the UK.

In 1981 the NEB's shares in Celltech were transferred to a new government agency, the British Technology Group (BTG), which had been formed by a merger between the NEB and the NRDC. While continuing the NRDC's licensing activities (obtaining technology from universities and public laboratories and licensing it to companies around the world), BTG took over the NEB's interests in small, high-technology firms.[16] These included participation with venture capitalists in Agricultural Genetics Company, set up in Cambridge to exploit technology coming out of the Agricultural Research Council's laboratories; it was known as Celltech's country cousin.[17]

---

[14] Searle was one of several US pharmaceutical companies which had established subsidiaries in the UK after the war to supply drugs to the National Health Service. Under the government's price regulation scheme companies that invested in UK research laboratories were entitled to higher prices than those which did not (see Chapter 8).

[15] Interview with John Berriman, 18 February 2014.

[16] BTG was privatized in 1992 through a management buy-out, and went public in 1995. It subsequently shifted away from licensing to focus almost entirely on the life sciences; this change of strategy is described in Chapter 4.

[17] *Financial Times*, 15 December 1981. Agricultural Genetics Company was not a financial success, partly because one of the government laboratories which had been expected to be an

In 1983, when new shares were issued to raise an additional £6m, ICFC sold its holding, apparently because it was unhappy with Celltech's management. Its shares were acquired by Biotechnology Investments Limited (BIL), an investment fund specializing in biotechnology which had been set up by N.M. Rothschild, the merchant bank; most of BIL's previous investments had been in US firms, including Amgen and Genzyme, and it was now looking for opportunities in Europe.[18] Lord Rothschild, chairman of the bank, had been trained as a biologist and had been head of research at Royal Dutch Shell; he was friendly with the LMB's Sydney Brenner, who acted as an adviser to BIL.

The Thatcher government's investment in Celltech was not part of a comprehensive plan for biotechnology; the concept of industrial strategy, with the government singling out particular industries for special support, had been discredited by the failure of Labour's interventionist policies in the 1960s and 1970s, and Mrs Thatcher was determined not to repeat them. Her prime concern was to inject a new spirit of enterprise into the economy by lowering taxes and by removing obstacles to the growth of all entrepreneurial firms, irrespective of the industry they were competing in.

This was the rationale for the Business Start-up Scheme, later renamed the Business Expansion Scheme, which enabled private individuals to claim tax relief of up to £40,000 on gains made from investments in unquoted companies, provided the shares were held for 5 years. The scheme was designed to encourage wealthy individuals and successful entrepreneurs to become business angels, investing in start-up firms, mentoring them, and helping them to grow. The government also altered the tax rules to make stock options more attractive. In tax policy, as in other areas, the government was seeking to create an enterprise culture comparable to that of the US.[19]

The Growth of Venture Capital

The remaking of the UK as an entrepreneur-friendly economy attracted investment from American venture capitalists, who saw an opportunity to apply the same investment techniques that had proved highly profitable in the US.

Among several new entrants to the venture capital business at the start of the 1980s was Advent, which began as a partnership between Peter Brooke,

---

important constituent of the company, the Plant Breeding Institute, was sold to Unilever in 1986. Several of the other businesses were sold to the private sector during the 1990s.

[18] BIL became a closed end investment trust, listed on the stock market, in 1984.

[19] Nigel Lawson, *The View from No 11: Memoirs of a Tory Radical* (London: Corgi Books 1993) p. 344.

managing partner at a Boston-based firm, TA Associates, which had invested in Biogen and other US biotechs, and David Cooksey, a British businessman.[20] Cooksey was to become a successful investor in biotech and an influential adviser to successive governments on policy towards the life sciences.[21]

The first three Advent funds, backed by Scottish investment trusts and insurance companies, were set up as Jersey-based limited companies, since UK tax rules did not permit venture capital firms to be organized as limited partnerships on the US pattern. These restrictions were removed by the Inland Revenue in 1985, and Advent, renamed Advent Venture Partners, raised subsequent funds as UK limited partnerships, as did other British venture capital firms. Each fund had a fixed life, usually 10 years, at the end of which the investors expected a return on their capital invested.

Another new firm with a US connection was Alan Patricof Associates, later renamed Apax Partners. This was the product of collaboration between an American venture capitalist, Alan Patricof, and Ronald Cohen in the UK.[22] Like Cooksey at Advent, Cohen recruited scientists as well as finance experts to join the firm, and some of them served on the boards of the companies in which Apax invested.

Among established British institutions, the ICFC began to build a venture capital business alongside its original role as a supplier of loan finance. In 1983 it was renamed 3i (short for Investors in Industry) and its venture capital arm, Technical Development Capital, became 3i Ventures. Relations between 3i and the new venture capital firms were cool. 3i did not join the British Venture Capital Association (BVCA) when it was formed, with Cooksey as chairman, in 1983, apparently because it did not accept the BVCA's insistence on 'hands-on' investing—a close involvement with the companies backed by venture capitalists—as a condition for membership.[23] However, it joined the association later in the decade and became an active investor in biotech start-ups. With its nation-wide network of offices it was well placed to identify promising businesses, and, thanks to the financial backing provided by its owners, it could take a long-term view of its investments.[24]

Several of the big investing institutions, including Prudential Assurance, Commercial Union, and Legal & General Assurance, created their own venture

---

[20] Peter A. Brooke, *A Vision for Venture Capital* (Boston: New Ventures Press, 2009).

[21] Before joining Advent, Cooksey had turned round a loss-making plastics and packaging business, Intercobra, which he had bought from De La Rue. Christopher Lorenz, 'Why David Cooksey went up market', *Financial Times*, 10 November 1981.

[22] Ronald Cohen, *The Second Bounce of the Ball* (London: Weidenfeld and Nicolson, 2007) pp. 23–4.

[23] *Financial Times*, 1 February 1983.

[24] 3i went public in 1994 and invested in a number of UK biotech firms. In 2008 it switched its attention away from venture capital to management buy-outs and what came to be called private equity transactions (*Financial Times*, 26 March 2008).

capital arms.[25] The most active of the merchant banks were Rothschild, through BIL, and Schroders; the latter's life sciences arm, SV Life Sciences, was established as a separate legal entity in 1993.

In 1987 one of Rothschild's investment directors, David Leathers, who had been involved in creating BIL and had managed its investments, broke away to join Abingworth, an independent venture capital firm in London which was seeking to extend its interests into biotech. Set up in 1973 by two London stockbrokers, Peter Dicks and Anthony Montagu, Abingworth had made some successful bets on US electronics firms such as Apple and Silicon Graphics, investing alongside US venture capitalists, and they hoped to do the same in biotech.

Leathers brought with him from Rothschild Stephen Bunting, who was to become a powerful force in UK biotech; he was appointed Abingworth's managing partner in 2002. Bunting had a scientific background, having taken his first degree in biological sciences at the University of Sussex and then a PhD in evolutionary genetics at Reading; he had worked as a venture capitalist for the Prudential before Leathers recruited him to join the BIL team at Rothschild.

By the mid-1990s Abingworth had established itself as one of the leading British venture capital firms in the biotech sector, often investing alongside Advent and Schroders. These three firms, together with 3i and Apax, became the principal suppliers of capital for start-up and early-stage biotech firms. The sector also attracted investment from other parts of Europe, including Atlas Venture, based in Holland, Sofinnova from France, and TVM from Germany. Margaret Thatcher's ambition to make the UK a land fit for entrepreneurs appeared to be bearing fruit.

New Entrants in Biotech

A more attractive tax regime for entrepreneurs, easier access to venture capital, admiration for what Genentech and other US firms were doing—all this contributed to a wave of biotech start-ups in the 1980s. Several of these UK firms were founded with the intention of entering the field of drug development—six between 1980 and 1984, nineteen between 1985 and 1989.[26]

---

[25] Prudential's venture capital subsidiary, Prutec, had links with the Patscentre, the science and technology arm of PA International, management consultants. Commercial Union and Legal & General set up a jointly owned venture company, Cogent. Vivien Lee, John Gurnsey, and Arthur Klausner, 'New trends in financing biotechnology', *Nature Biotechnology* 1, 7 (September 1983).

[26] Michael M. Hopkins, Philippa A. Crane, Paul Nightingale, and Charles Baden-Fuller, 'Buying big into biotech: scale, financing and the dynamics of UK biotech, 1980–2009', *Industrial and Corporate Change*, 22, 4 (August 2013) pp. 903–52.

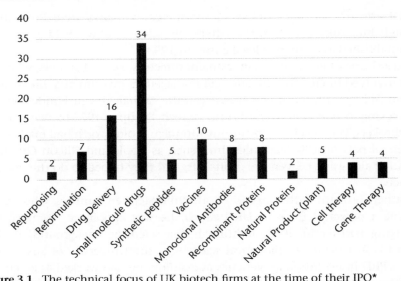

Figure 3.1. The technical focus of UK biotech firms at the time of their IPO*

* Firms are recorded as active in multiple categories, in keeping with self-reporting in the IPO prospectus.

Source: Science Policy Research Unit, University of Sussex.

Only a few of them were 'pure' biotechnology firms, in the sense that they relied on recombinant DNA or monoclonal antibody technology. Indeed it is notable that of all the emerging UK therapeutic biotech firms, only Celltech took a licence to use Cohen and Boyer's recombinant DNA technology (widely licensed by US biotechs and other more established firms around the world).[27] The low engagement of the UK sector with the novel biotechnology techniques is illustrated by Figure 3.1, which shows the technical focus of the sixty-six UK biotech firms that had an IPO before 2010 (based on the pipeline of planned projects they describe in their prospectuses).[28] The figure shows that the majority of these firms were either 'mini-pharmaceutical' firms, developing small molecule drugs, or drug delivery firms (e.g. making sustained release or inhalable formulations of previously approved drugs). Although new biotechnologies continued to emerge, such as gene therapy and stem cell therapy, firms using more traditional therapeutic strategies, based on natural products (such as botanical plant extracts) or vaccines, continued to be more

[27] The Cohen–Boyer patents were licensed to over 450 organizations globally, but Celltech was the only UK therapeutic biotech firm to take a licence. (Personal communication based on Stanford licensing data—Andrew Nelson, University of Oregon, July 2015.)

[28] Firms can have multiple technologies, hence the numbers do not sum. Although these sixty-six firms account for just over a quarter of the UK biotech firms focusing on therapeutic R&D, they obtained disproportionately more funding, due to their access to stock market investors. These firms formed the core of the emerging UK biotech sector—together with eleven firms that joined the stock market without IPOs (not included).

numerous on the UK's stock exchanges. Nonetheless, whatever the composition of their pipelines, all of them came to be regarded by investors and commentators as part of the emerging biotech sector.

The founders of the new firms fell into three categories: non-scientist entrepreneurs who saw biotech as a money-making opportunity; scientists with an entrepreneurial bent who were working in the pharmaceutical industry or for US biotech firms; and academic scientists based in universities, teaching hospitals, or government laboratories. Academic scientists were the smallest of the three categories in the early years, reflecting an important difference between the UK and the US, the limited involvement of British universities in the early biotech ventures. This explains, to some extent, the relatively low engagement of the sector with novel techniques emerging from the university research community.

Prominent in the first group was Wensley Haydon-Baillie, who had run an optical instruments company and had spent 2 years with Slater Walker, an investment bank that rose and fell in the 1960s and 1970s. His company, Porton International, had a set of drug programmes which included marketing rights to a herpes vaccine developed at Birmingham University. Haydon-Baillie also secured exclusive access, against competition from established pharmaceutical groups, to technology coming out of the Centre for Applied Microbiology and Research (CAMR) Laboratory at Porton Down in Wiltshire.[29]

Haydon-Baillie told journalists that his long-term ambition was for Porton to grow as big as Glaxo.[30] He was a persuasive salesman; to an extent unmatched by any other new entrant he convinced institutional investors to invest directly in his company, raising £15m from a group of investors that included Sun Life and Legal & General as well as the pension funds of Barclays Bank, Esso, and ICI.[31] Porton's credibility was improved in 1988 when he recruited John Burke, a senior Glaxo executive, to be chief operating officer.

Australian-born Ian Gowrie-Smith had a background in investment management. In 1986 he set up a drug development company, Medirace, and floated it a year later on the London Stock Exchange's Third Market.[32] Medirace worked on a novel cancer drug (Contracan) which had been developed by scientists at Hammersmith Hospital, but Gowrie-Smith was also looking for acquisitions. In 1990 he bought Evans Medical, which owned a vaccine plant at Speke, near Liverpool. The name of the company was changed to Medeva and Gowrie-Smith brought in Bernard Taylor from Glaxo to be chief

---

[29] David Fishlock, 'A £15m landmark for Porton', *Financial Times*, 18 October 1985.
[30] *Financial Times*, 4 December 1988.
[31] Peter Marsh, 'Rich promise of herpes vaccine delayed', *Financial Times*, 30 June 1988.
[32] The Stock Exchange had set up the Unlisted Securities Market in 1981 to cater for small firms that could not meet the listing conditions of the main market. The Third Market was launched in 1987 to accommodate even smaller firms.

executive of what was now seen as a potentially high-growth pharmaceutical company.[33]

Gowrie-Smith left the company in 1993 (he went on to create a drug delivery firm, SkyePharma), but Medeva continued to grow through acquisitions, of which the largest was the $400m purchase of a portfolio of drugs from Rhône-Poulenc Rorer in 1996; the portfolio included a range of over-the-counter cough medicines as well as several prescription drugs.

In the second category—founders coming out of the pharmaceutical industry—the best-financed and for a time most highly regarded new entrant was British Biotechnology. The founders were Brian Richards and Keith McCullagh, who, like Carey at Celltech, had been working for the UK subsidiary of the American pharmaceutical company, Searle. In 1985 Searle was taken over by Monsanto, and in the following year the new owners closed down the UK research laboratories, which were thought to duplicate the work of its St Louis facilities. Richards and McCullagh then set up on their own, recruiting a team of ex-Searle scientists to form the new company.

McCullagh was a qualified veterinary surgeon with a PhD in comparative pathology (he had spent some time in Africa, studying cardiovascular disease in elephants) and he worked for 4 years as a post-doctoral researcher at the Cleveland clinic in the US, a leading centre in heart surgery. He returned to the UK in 1974 to take up an academic post at Bristol University, before joining Searle in 1980. During his time at Searle, as he said later, 'I had watched the creation of Genentech and then its amazingly successful flotation, and then Amgen and Biogen—we had visited many of them in Searle to discuss collaboration. We thought, we have all the skills to do this, why can't we create something like that in the UK? At the time there was only one British biotech company of note, Celltech, and I thought the Searle team could create something different.'[34]

McCullagh was a charismatic leader, capable of enthusing colleagues and potential investors alike, and the new company had an American feel to it. Its lead investor was Rothschild's BIL. Jeremy Curnock-Cook, who had taken over as head of BIL after David Leathers had gone to Abingworth, commented: 'They were exactly the sort of people we wanted to back. They had strong leadership, good instincts and a strong sense of where they were going.'[35] Other investors included the venture capital arms of two US pharmaceutical groups, Johnson & Johnson Development Corporation and Smith Kline Beckman's SR One; their presence gave added credibility to the new company.

---

[33] Richard Gourlay, 'Building a dream on drug marketing', *Financial Times*, 5 March 1992.

[34] Interview with Keith McCullagh, 11 December 2013.

[35] Jeremy Curnock-Cook, Biotechnology Investments Ltd, quoted in *Sunday Times*, 26 May 1996.

Another entrepreneurial scientist who was influenced by the American model, and became a high-profile advocate for the biotech sector, was Chris Evans. A Welshman with a strong attachment to the country of his birth, Evans had studied microbiology at Imperial College, followed by a PhD at the University of Hull and a research fellowship at the University of Michigan.[36] After a spell working for Allelix, a Canadian biotech, he returned to the UK in 1986 as head of R&D in the British subsidiary of Genzyme. Evans was energetic and ambitious, eager to strike out on his own. In 1987 he left Genzyme with several colleagues to form his own company, Enzymatix, based in Cambridge (but not directly connected to the university).

The funding for Enzymatix came from S & W Berisford, a diversified group which owned British Sugar, the principal UK sugar refiner. Evans was a microbiologist and an expert in enzymes, and Berisford's idea was that biocatalysis could be used to generate high-value products with the waste material from sugar refining.[37] Berisford ran into a financial crisis in 1990 and sold British Sugar, but continued to own Enzymatix until 1992, when Evans and his associates acquired control of the company.

A more modest start-up, but one that ultimately became the most successful of all the UK biotech firms started in the 1980s, was Shire Pharmaceuticals. The principal founder was Harry Stratford, who had spent most of his career with Wyeth-Ayerst, a subsidiary of the US pharmaceutical company, American Home Products; he had been responsible for the marketing of Premarin, a hormone replacement therapy. Keen to start his own business, he joined with four colleagues to set up Shire in 1986.[38] The starting-point was a range of calcium products for treating bone disease, licensed from Nycomed in Norway. Backed at the start by Schroders, Shire based its growth on licensing drugs or drug candidates from other companies and then marketing them—a 'search and develop' strategy, as distinct from research and development.

By contrast, academic scientists—the third group of company founders—aimed to use innovative science as the basis for developing new treatments for disease. A leading member of this group was Greg Winter at the MRC's

[36] In the offer document for Chiroscience, one of Evans's companies which went public in 1994, he was described as a leading authority in biocatalysis and chiral chemistry, with over 100 scientific publications and patent applications to his credit.

[37] Biocatalysis is the use of biological molecules such as protein enzymes to catalyse chemical reactions. Biocatalysis can speed up reactions as well as reducing the energy needed to complete reactions.

[38] Shire's original name was Aimcane; it was based in Andover, Hampshire, and later established a development laboratory at the nearby town of Basingstoke. In addition to Stratford, the other co-founders were Jim Murray, an expert in osteoporosis and women's health; Dennis Stephens, a consultant; Peter Moriarty, who was then working for Schering-Plough; and John Kanis, professor of human metabolism at the University of Sheffield. Shire plc, *Flying high, the first twenty years at Shire Pharmaceuticals* (Shire plc 2006).

Laboratory of Molecular Biology in Cambridge. The MRC, after helping to set up Celltech in 1980, had taken steps to strengthen its links with other companies, setting up a technology transfer arm, later renamed MRC Technology, in 1984. A particular focus was on monoclonal antibodies, building on the discoveries made by Milstein and Köhler. The application of this technology in medicine had been more difficult than expected, partly because the murine content (produced from mouse cell culture) in the early antibody-based drugs provoked an immunogenic response in patients.

Winter had developed a technique for making humanized antibodies, greatly reducing their murine content. The MRC licensed this technology on a non-exclusive basis to pharmaceutical and biotech companies around the world, contributing to what was to become the fastest-growing segment of the biologics market.[39] Winter's next step was to develop fully human antibodies, based on phage display technology, and this opened up another promising route towards effective antibody-based therapeutics.[40] In this case, because the technology was still at an early stage and needed several years before it could be fully commercialized, Winter agreed with the MRC that the technology should be licensed to a start-up firm.[41] This firm would be totally devoted to the project and financed by venture capitalists and other investors.[42]

Winter joined forces with David Chiswell, a molecular biologist who had been working for Amersham, the diagnostics group, and the two men founded Cambridge Antibody Technology (CAT) in 1989.[43] Seed funding came from Peptide Technology (Peptech), an Australian biotech whose founder and chairman, Geoffrey Grigg, had worked at the LMB in the 1970s and had kept in close touch with Winter. Peptech was CAT's principal shareholder until 1996, shortly before the company went public. The MRC also had a small shareholding in the business.

In his research into monoclonal antibodies Winter collaborated with scientists in Cambridge University, notably Herman Waldmann in the Department of Pathology, but at this stage neither Cambridge nor any other British university was actively engaged in promoting spin-off firms. In 1985 the Thatcher

[39] By 2015 patents on the MRC's intellectual property in monoclonal antibodies had generated royalty income of about £600m. Income from this source was expected to decline over the next few years as the patents expired. *Financial Times*, 17 August 2015.

[40] Phage display became one of the two leading platforms for antibody-based drug development. The other, based on transgenic mice, was developed by two US companies, Medarex and Abgenix.

[41] The debates within the MRC that preceded the establishment of Cambridge Antibody Technology are described in Soraya de Chadarevian, 'The making of an entrepreneurial science, biotechnology in Britain 1975–1995', *Isis*, 202, 4 (December 2011) pp. 601–33.

[42] Winter had some experience of start-ups, having advised Scotgen, an antibody engineering firm founded in 1987 by Bill Harris, professor of genetics at Aberdeen University.

[43] Amersham had originally been the radiochemical centre of the Atomic Energy Authority; privatized in 1982, it specialized in diagnostic devices, and became an important source of management talent for British biotech firms.

government had abolished the monopoly which the NRDC, now British Technology Group, had exercised over the exploitation of government-funded research; the universities were allowed to keep title to the intellectual property created by their scientists. The aim, as Keith Joseph, Secretary of State for Education and Science, explained, was to 'maximise the chances that good inventions would be identified, assessed, protected and exploited'.[44]

The change brought British universities into line with their US counterparts after the passage of the Bayh–Dole Act (see Chapter 2), but they were not well equipped to take advantage of their new freedom. The first moves to create new firms came from individual scientists who received little if any support from the university authorities.[45]

Alan Munro, head of immunology in the Department of Pathology in Cambridge, was unusual among academics in having an interest in business. In the 1987/1988 academic year he took sabbatical leave to work in project management at Celltech. He was then asked by Stephen Bunting to do consulting work for Abingworth, which had just set up its biotech arm and was looking for projects. Out of that assignment came the idea for a new company, first called Immunology Ltd and later renamed Cantab Pharmaceuticals. Its first drug candidate, aimed at reducing the incidence of organ rejection in kidney transplants, was based on a monoclonal antibody developed by Munro's Cambridge colleague, Herman Waldmann. That project was later dropped and Cantab focused on therapeutic vaccines to treat cancer and persistent viral infections.

Abingworth provided seed funding for the new firm (other VCs came in later) and Bunting was a co-founder, together with Munro and Andrew Sandham, who had worked with Munro in Celltech.[46] Munro left the university to become chief scientific officer, a post which he held until 1995. The university itself played no role in the creation of Cantab.

Oxford was even less interested than Cambridge in commercialization, and the first university-linked biotech firm came about through an unusual route. The two principal players were Monsanto, the US chemical company, and Raymond Dwek, a professor in Oxford's Department of Biochemistry. Monsanto, which was then scouring the world for promising biotechnology-related research, became aware of Dwek's work on sugars that were attached to

---

[44] *Financial Times*, 15 May 1985.

[45] Universities were not the only source of spin-outs. Two professors or ex-professors who founded biotech firms during the 1980s had been working in hospitals. Dennis Chapman from St Bartholomew's Hospital co-founded Biocompatibles in 1984, and John Landon from the Royal Free Hospital founded Polyclonal Antibodies in 1983 and Therapeutic Antibodies in the following year; these two firms were merged in 1990.

[46] Sandham, who had started his career at Amersham, was later involved in numerous biotech firms as founder, chief executive, or director. Alan Munro served as chief scientific officer of Cantab until 1995, when he was appointed Master of Christ's College, Cambridge.

proteins and lipids, the field that was later given the name of glycobiology (invented by Dwek and given the stamp of approval by the *Oxford English Dictionary*).[47] The American company saw that this work had commercial potential, and in 1985 they offered a 5-year grant to support Dwek's research.

This was an unfamiliar proposition for Oxford academics, and questions were raised about whether it was right to accept industrial money; some feared that a Monsanto grant might distort the university's research priorities. Thanks to Dwek's persistence and Monsanto's skilful diplomacy the opposition was overcome; it was the first industrial grant in Oxford's 850-year history, and amounted in total to about $100m over a 13-year period. Further cooperation between Dwek (now head of a new Institute of Glycobiology), Monsanto, and its UK subsidiary Searle led to the creation in 1988 of Oxford GlycoSciences (OGS), based on the research that Monsanto had supported. Dwek went on the board, but retained his university position; his academic colleague and co-founder of OGS, Raj Parekh, joined the company as research director.

The company was started as a supplier of enzymes, instrumentation, and other products for use in carbohydrate-related research, but it switched to drug discovery and developed what became its lead drug—miglustat (Zavesca)—an oral therapy for Gaucher's disease. Dwek was determined from the start that OGS should become a world-class company, and that could not be done without financial backing from the US. The initial investors were Advent Capital and Euro Ventures (both managed by Advent) and Alafi Capital Corporation from the US; Monsanto was a limited partner in each of these funds. The first two chief executives were Americans who had held senior posts in US biotechnology firms. Oxford University was also an investor—the first time it had been directly involved in a spin-off firm.

Margaret Thatcher, who visited Dwek's institute several times, saw Oxford GlycoSciences as a model for British science—fundamental research in the university partly funded by industry, leading to a spin-off company.[48]

The Stock Market Opens Up

By the end of the 1980s a British biotech sector was starting to take shape, but progress had been slower than expected. Celltech, in particular, was coming under criticism for its lack of clear commercial direction. While it was

---

[47] Glycobiology is defined in the *Oxford English Dictionary* as the scientific study of carbohydrates and their role in biology.

[48] Raymond A. Dwek, 'Journeys in science: Glycobiology and other paths', *Annual Review of Biochemistry*, 83 (2014) pp. 1–44.

generating revenue from making monoclonal antibodies for sale to other companies and from other contract services, it had been slow to identify and develop its own drug candidates.

To correct these weaknesses John Jackson, Cellech's chairman, engineered the appointment of Peter Fellner, who had been managing director of Roche's UK subsidiary, to succeed Gerard Fairtlough as chief executive.[49] Fellner had started his career as an academic scientist, specializing in molecular biology; like many of the leading figures in UK biotech he had worked for several years at Searle before joining Roche in 1984.

Just before Fellner took over, the ownership of Celltech had been put in doubt when its largest investor, British and Commonwealth, which was in the throes of a financial crisis, put its 36 per cent shareholding up for sale. If, as seemed likely, the shares were sold to a single buyer, this might have triggered a bid for the whole company. There was speculation that one of the British pharmaceutical companies might buy Celltech, but by this time they were developing their own biotechnology capabilities, and showed little interest.[50] Sir Alfred Shepperd, chairman of Wellcome, said: 'Why should we want Celltech? We have got a biotech group already.'[51] Among non-British companies, Bayer from Germany was interested in Celltech's manufacturing business, but after inspecting the Slough facilities decided not to proceed.[52] After British and Commonwealth went bankrupt, the shares passed into the hands of the company's administrators, who retained them until Celltech went public in 1993.

The brightest star at this point was British Biotechnology. This company, according to one commentator, 'appears to have all the ingredients investors are seeking in biotechnology in the late 1980s, including novel ideas, technical talent and experienced management'.[53] Richards and McCullagh were building up their research staff and at the same time strengthening their relations with the City. In 1989 they raised £22.7m in a private placing, bringing in twelve British and three Japanese financial institutions as investors. The placing was

---

[49] Jackson had been on Celltech's board from the start as BTG's representative; he took over the chairmanship in 1983.

[50] The attitude of the British pharmaceutical companies towards Celltech had been sceptical or hostile from the start. Most of Celltech's collaboration agreements had been with non-British firms. Dodgson, *Celltech*, p. 85.

[51] Peter Marsh, 'When a sweet pill is spiked by bad timing', *Financial Times*, 30 January 1990. Wellcome was the most active of the British firms in biotechnology during the 1980s. Its biotechnology subsidiary was closed down in 1991 as part of a cost cutting drive by a new management (*Financial Times*, 3 May 1991). Glaxo, the largest British pharmaceutical company, had taken several steps to establish a position in biotechnology, including the purchase in 1987 of Biogen's Geneva laboratory, *Financial Times*, 29 July 1987.

[52] Bayer was interested in Celltech's antibody production facilities, but decided that they were not big enough.

[53] David Fishlock, 'Six companies with an eye on the future', *Financial Times*, 27 May 1988.

handled by Kleinwort Benson, a leading merchant bank; one of the bank's managers, James Noble, was recruited to join British Biotechnology as finance director at the end of 1990, in time to help the company raise a further £40m to fund product development through a private placement.[54]

If British Biotechnology was to continue to grow, it would need access to public market investors, just as Genentech and its imitators had done in the US during the 1980s. However, the London Stock Exchange did not allow loss-making firms to obtain a full listing, and although the Exchange had launched a junior exchange, the Unlisted Securities Market, in 1981, the amounts of money that could be raised were too small to attract more than a handful of biotech firms.[55] An alternative for British firms was to go to NASDAQ, but as Cantab found when it listed its shares on NASDAQ in 1992 it was hard to interest US investors in a small, loss-making British firm with no operations in the US.[56]

British Biotechnology, which had already raised substantial sums from international investors, was a more promising candidate for NASDAQ, and the threat that it might choose New York rather than London persuaded the Exchange to make an exception in this case, resorting to a rule that had been used in the Eurotunnel flotation of 1987. This rule allowed a loss-making company to float if the exchange was satisfied that potential investors had enough information about the business on which to base a decision; Euro-tunnel when it floated had no prospect of making profits before 2000.[57]

British Biotechnology chose to list in London and New York on the same day, 9 July 1992—the first time this had been done by any British company—although the bulk of the money came from UK investors. It raised about £30m, expanding its share capital by about 20 per cent; its market capitalization after the flotation was about £150m and it stayed at that level for the next 2 years. The name of the company was later shortened to British Biotech.

The Exchange then came under pressure from other biotech firms and their brokers to go further. In 1993 it introduced new rules that allowed loss-making biotech firms to obtain a listing as long as they met certain conditions.[58] They had to demonstrate their ability to attract funds from sophisticated investors; they had to have a market value of at least £20m, a 3-year record of operations

[54] Noble is one of several ex-accountants who have played a large role in the UK biotech sector. He qualified as a chartered accountant with Price Waterhouse in 1983.

[55] This was also true of the Third Market which was set up in 1987. Only two therapeutic biotechnology firms joined the USM. The Third Market was merged with the USM in 1990.

[56] Two other firms that listed on NASDAQ were Ethical Pharmaceuticals (later renamed Amarin) and Xenova; the latter later listed in London.

[57] Clive Cookson and Daniel Green, 'The appliance of science in New York and London', *Financial Times*, 12 June 1992.

[58] These conditions and their rationale are set out in a memorandum submitted by the London Stock Exchange to the House of Commons Select Committee on Science and Technology, 28 April 1999.

in research and development, and two drugs in clinical trials. Keith McCullagh of British Biotech described the rule change as marking a 'coming of age in the biotech scene in Europe', a step towards forming a more US-style innovation ecosystem.[59]

To make themselves attractive to investors biotech firms needed a credible management team, a plausible plan for bringing drugs to the market, and preferably also a demonstrated ability to negotiate partnerships with Big Pharma. Such partnerships would not only generate revenue (and make it less necessary to ask shareholders for more money), but also provide some reassurance that their technology was valuable.

At Celltech, which floated at the end of 1993, Peter Fellner had decided soon after taking over as chief executive to separate the two main businesses, contract manufacturing and drug discovery. On the discovery side, there was now a clearer focus on three therapeutic areas: inflammatory diseases, including asthma and septic shock; autoimmune diseases, including rheumatoid arthritis; and certain types of cancer. Shortly before the flotation Fellner concluded a potentially lucrative partnership with Bayer on a treatment for Crohn's disease (CDP571).[60] Celltech already had partnerships with other pharmaceutical companies, including one with Lederle in the US for the development of a monoclonal antibody for the treatment of leukaemia.

For Chris Evans, the opening-up of the stock market created an opportunity to split his company, Enzymatix, into several separate businesses and to prepare them for flotation or for a trade sale. Evans had recruited as his chief financial officer Peter Keen, an accountant who had worked in Cambridge for Agricultural Genetics and then for a diagnostics company, Cambridge Life Sciences. The two men had very different personalities, but they formed an effective partnership, with Keen acting as a foil to Evans's enthusiasm while also sharing the same entrepreneurial ambition.

Evans attracted a group of talented managers, several of whom later served as chief executives in other biotech firms. Andy Richards, who joined Enzymatix in 1991 after previously working for ICI and PA Consulting, was an expert in chiral compounds; he later became a serial entrepreneur in biotech.[61] Edwin Moses, who was recruited as business development manager and later became

---

[59] Quoted in *European Biotech 1994*, Ernst & Young's first annual report on the European biotechnology industry.

[60] This was a humanized monoclonal antibody, one of a number of anti-TNF (tumour necrosis factor) drugs that would compete for a multi-billion dollar market in future years. Crohn's disease is a chronic inflammatory disease of the intestines.

[61] Chiral molecules are those that share the same chemical composition but are structurally distinct, occurring in different forms, or isomers, comparable to the difference between left and right hands. Chirality in drug formulation reduces efficiency of production and risks exposing patients to molecules that are potentially less therapeutically active or may cause side effects; reducing chirality to produce single isomer drugs may increase drug potency.

commercial director, had worked for a US biotech firm and then for Amersham. He later headed an Oxford-based start-up, Oxford Asymmetry, before moving to Ablynx in Belgium as chief executive.[62]

The most valuable part of Enzymatix was Chiros, a drug development business based on chiral technology; the name was later changed to Chiroscience to avoid confusion with Chiron in the US. Its most promising drug was Chirocaine, a local anaesthetic. Chiroscience was bought out from Enzymatix in 1992 for £1m and when it was floated on the stock market in 1994 it achieved a market value of £102m, ultimately producing a remarkably profitable exit for the three venture capital firms—3i, Schroders, and Apax—which had backed it, as well as for the founders.[63]

At Chiroscience Evans had recruited as chief executive Nowell Stebbing, who after several years with Searle in the UK had held research posts in the US with Genentech and Amgen. Shortly before the flotation Stebbing was diagnosed with cancer and although he continued as deputy chairman he was replaced as chief executive by John Padfield, a senior executive in Glaxo.

As with Fellner at Celltech, the presence of an ex-Big Pharma manager at the top of a biotech firm was comforting for investors, even though these recruits, accustomed to working in large, process-driven organizations, did not always adjust easily to the free-wheeling culture of a start-up firm. Many of the biotech firms that came to the stock market in the 1990s and later recruited their senior managers from Glaxo (GlaxoWellcome after the 1995 merger, and later GlaxoSmithKline) and AstraZeneca, although Searle was also an important source (Figure 3.2).

A Big Pharma appointment that worked well took place at Shire, which went public in 1996. Two years before the flotation Harry Stratford retired as managing director—he was keen to start another new company[64]—and was replaced by Rolf Stahel, a Swiss national who had spent 27 years with Wellcome, most recently as director of group marketing.

Stahel was not impressed by Shire's product range, but the marketing organization was strong and he decided that running a new firm might be more interesting than staying in Big Pharma.[65] He hired as his chief financial officer an investment banker from Lazard Brothers, Stephen Stamp, and the two men developed Shire's distinctive strategy, which depended mainly on

---

[62] The main attraction of Enzymatix, Moses recalled later, was working with 'the incredibly inventive and hard-working Chris Evans'. Interview with Edwin Moses, 22 May 2013.

[63] According to the *Financial Times*, analysts had expected a valuation of about £80m. However, the newspaper noted that Chiroscience was one of the few companies developing single isomer drugs which, if they fulfilled their promise, could revolutionize parts of the pharmaceutical industry. *Financial Times*, 28 January 1984.

[64] Stratford set up Strakan (renamed ProStrakan after a merger with a French firm) in 1995; it had a similar business strategy to Shire.

[65] Shire plc, *Flying high*, p. 49.

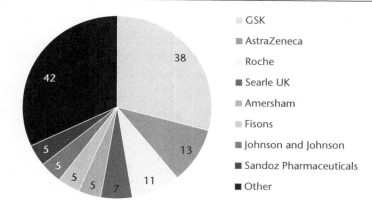

Figure 3.2. Number of Chairmen, Chief Executive Officers, and Chief Scientific Officers recruited from pharma companies into biotech at the time of their IPO
Source: Science Policy Research Unit, University of Sussex.

in-licensing and acquisitions. As they told investors at the time of the IPO, Shire did not engage in early-stage research or seek to exploit a unique technology.

Even better than executives from Big Pharma, from the investors' point of view, was the presence of top managers who had worked for biotech firms in the US. Oxford GlycoSciences, which went public in 1998 at a time when investor enthusiasm for biotech was cooling, was unusual in having two American executives in senior posts. Kirk Raab, who had worked under Swanson at Genentech and served as chief executive for several years after the Roche deal in 1990, was non-executive chairman; Michael Kranda, who had spent 11 years with Immunex, another first-generation biotech firm, was chief executive from 1996 to 2002.

A total of nineteen biotech firms focusing on drug discovery listed on the London Stock Exchange main market in the 1990s (Table 3.1). Until the mood changed towards the end of the decade, the reaction of investors was generally positive. In 1996 the stock market value of the UK biotech sector was close to £4bn, about the same as the merchant banking sector.[66] When PPL Therapeutics, the Edinburgh-based firm which was to become famous for its commercialization of cloned sheep to produce therapeutic proteins (following its founder's role in creating Dolly the sheep), raised £30m in that year, it could have raised more, but was told by its financial advisers not to be too greedy.[67]

---

[66] *Financial Times*, 8 July 1996.   [67] Private communication.

Table 3.1. Biotech IPOs on the London Stock Exchange's main market, 1992–1999

Company	Year	Amount raised (£m)	Market capitalization at IPO (£m)
British Biotech	1992	30.0	151.8
Celltech	1993	50.0	176.5
Scotia	1993	40.6	181.0
Cantab	1993	13.8	45.3
Chiroscience	1993	45.0	102.0
Cortecs	1994	10.6	47.5
Biocompatibles	1995	15.1	74.0
Peptide	1995	24.0	68.0
Shire	1996	40.0	106.6
Xenova	1996	22.6	50.0
Vanguard Medica	1996	46.5	111.5
Phytopharm	1996	11.3	54.1
PPL Therapeutics	1996	30.0	101.3
Cambridge Antibody	1997	41.0	109.0
PowderJect	1997	35.0	110.0
Oxford GlycoSciences	1998	.30.1	103.2
Bioglan	1998	20.1	193.0
Quadrant	1998	20.0	57.7
Antisoma*	1999	–	74.0

* Antisoma, having listed on EASDAQ in 1998, listed in London at the end of 1999 but did not raise new money at that time.

Source: London Stock Exchange, *Financial Times*.

The Introduction of AIM

In 1995 the Stock Exchange launched a new junior market, the Alternative Investment Market (AIM), to replace the Unlisted Securities Market.[68] This was not directly related to the needs of the biotech sector, but was the result of a long debate in the City about how best the exchange could serve the needs of small firms.[69] AIM was based on the principle of 'light touch' regulation, with less prescriptive rules that those of the main market: there was no requirement for a trading record, or for a minimum percentage of the equity to be in public hands. Whereas on the main market companies had to have their prospectus vetted by the Listing Authority (which at that time was the Stock Exchange), on AIM the responsibility for assessing the suitability of the companies rested with nominated advisers or Nomads—normally investment bankers, brokers, or accounting firms.

One of the first biotech firms to join AIM was a gene therapy company, Oxford BioMedica, set up in 1995 by a husband and wife team, Alan and Sue

[68] Sridhar Arcot, Julia Black, and Geoffrey Owen, *From local to global, the role of AIM as a stock market for growing companies* (London Stock Exchange, September 2007).
[69] Ranald Michie, *The London Stock Exchange, A History* (Oxford: OUP, 1999) pp. 618–22.

Kingsman, both of whom were scientists in Oxford University's biochemistry department.[70] (Alan Kingsman had previously served as associate director of research at British Biotech.) They raised £5m on AIM in 1996 and although the company's shares performed poorly in the months following the flotation, it raised £24m in secondary offerings before transferring to the main market in 2001. Six other biotech firms floated on AIM between 1995 and 1999.

Some venture capitalists, led by Ronald Cohen of Apax, criticized the decision to set up AIM.[71] A better approach, in Cohen's view, would be to create a pan-European exchange modelled on NASDAQ which would accommodate all Europe's leading high-technology firms; this would give investors a wider choice than was available on national exchanges. In 1995 Cohen and a few like-minded colleagues established a new exchange, EASDAQ, based in Brussels, with listing rules similar to those of NASDAQ. It was seen by the European Commission as a promising step towards the creation of an integrated European capital market.[72] However, EASDAQ was unable to generate much support either from investors or from firms planning to go public; most of them preferred to list on their local exchanges, several of which, like London, set up junior markets of their own.[73]

Only one British biotech firm listed on EASDAQ; Antisoma, a cancer specialist supported by the Imperial Cancer Research Fund (ICRF), did so in 1998, but switched to London at the end of 1999. There was sufficient appetite for biotech stocks in London, both through the main market and through AIM, to generate a sizeable flow of IPOs in the mid-to-late 1990s.

New Sources of Capital

The opening of the stock market and the favourable response from investors prompted the creation of new venture capital firms as well as investment trusts specializing in biotech. Chris Evans and Peter Keen, having successfully floated Chiroscience, set up an unusual two-headed organization. One part was Merlin Ventures, which would identify promising opportunities, principally in universities, and provide the scientists with seed capital to get started. The other was the Merlin Fund, which would offer larger amounts of funding and bring in other investors to support the businesses created by Merlin

---

[70] The Kingsmans were specialists in the area of gene expression and retrovirus biology and were responsible for numerous patents.

[71] Cohen, *Second Bounce of the Ball*, pp. 28–9.

[72] Dana T. Ackerly, 'Easdaq—the European market for the next hundred years?' *Journal of International Banking Law* 12, 3 (1987) pp. 86–91.

[73] These included, in addition to AIM in London, the Neuer Markt in Frankfurt and the Nouveau Marché in Paris.

Ventures. Several of the firms backed by Merlin were spin-outs from universities, including Cyclacel from Dundee, based on a new approach to tumour suppression invented by Professor David Lane, and Microscience, a vaccine firm co-founded by David Holden at Imperial College.[74]

Rivalling Evans as a founder of biotech firms was Alan Goodman, a non-scientist whose first job in the sector was as finance director in Agricultural Genetics. He worked with Gowrie-Smith at Medeva, with Evans in the early days of Chiroscience, and with the Kingsmans when they were setting up Oxford BioMedica. In 1982, together with Daniel Roach, Goodman formed a consulting business in Cambridge, Advanced Technology Management, which later became a fully-fledged venture capital firm, Avlar Bioventures.[75] Among the numerous firms which Goodman founded or co-founded was Peptide Therapeutics, a vaccine firm later renamed Acambis, which floated on the stock market in 1995.

A further addition to the supply of venture capital came from the Medical Research Council. David Owen, who had been recruited from SmithKline-Beecham to strengthen the MRC's links with industry, had found interaction with venture capital firms to be inefficient and erratic. Having tried and failed to form a partnership with 3i, he won approval from the MRC to create a new fund, the Medical Venture Fund, managed by Medical Venture Management—later renamed MVM Life Science Partners. Led by Stephen Reeders, who had a background in biomedical research and experience of early-stage healthcare investments in the US.[76] MVM raised its capital from the private sector, including pharmaceutical companies and financial institutions. Pfizer was an early investor, and its backing served to enhance MVM's credibility in the City. MVM's initial focus was on discoveries coming out of MRC-funded research—it had the right of first refusal on such discoveries before they could be offered to other firms—but the link with the MRC was subsequently loosened, and MVM became an independent venture capital firm, with offices in Boston and London.

---

[74] David Lane discovered the p53 gene, which in its mutant form is associated with many types of cancer. Its discovery was hailed as a major breakthrough. He was knighted in 2000. See Sue Armstrong, *P53: The Gene that Cracked the Cancer Code* (London: Bloomsbury Sigma, 2013).

[75] In the case of Peptide, before becoming chief executive, Goodman negotiated the acquisition from British Technology Group of some of the intellectual property on which Peptide was based; BTG became a substantial shareholder in the new company. BTG itself, having been privatized in 1992 by means of a management buy-out, was floated on the stock market in 1995; at this stage it was still primarily a technology transfer and licensing business, with interests in electronics and engineering as well as bioscience.

[76] Reeders had a doctorate in medicine from Oxford, and worked as a medical doctor for several years, before taking up research in molecular genetics, both in the US and in the UK. One of MVM's first investments was in Gendaq, a gene regulation company co-founded by Yen Choo, a post-doctoral researcher at the LMB, and Aaron Klug, who later took over from Sydney Brenner as director of the LMB. Gendaq was sold to Sangamo of the US in 2000.

Other sources of funding for start-up firms were the medical charities, of which the largest was the Wellcome Trust. Set up by Sir Henry Wellcome, the trust had been the sole shareholder in the pharmaceutical company, Wellcome Foundation (formerly Burroughs Wellcome and later renamed Wellcome Ltd) and it used its income to support medical research. The trust's resources were greatly increased after Wellcome Foundation became a public company and then, in 1995, when it merged with Glaxo. The trust sold most of its shares in the merged group, and its assets increased from £5.3bn to £6.8bn; by 2000 further diversification of the portfolio raised total assets to £15bn, making Wellcome the world's largest grant awarding charity. Most of the trust's income was used to support academic research, but it also invested alongside venture capitalists in biotech firms. One of its first investments, in 1997, was in Oxagen, a genetics firm spun out of the Centre for Human Genetics at Oxford, which was itself backed by the Wellcome Trust.

By the mid-1990s two of the UK's largest cancer charities, ICRF and Cancer Research Campaign, were also supporting biotech start-ups. In 1996 the ICRF took a minority stake in Antisoma, co-founded by one of its scientists, Agamemnon Epenetos, and in the same year the Cancer Research Campaign provided seed funding for Cyclacel in Dundee. The two charities merged in 2002 to form Cancer Research UK; its commercialization arm was called Cancer Research Technology (CRT).

To cater for retail investors, biotech-focused investment trusts were formed. In 1994 Rothschild established the International Biotechnology Trust, which had a narrower target than BIL. Its remit was to take larger stakes in fewer companies, most of which already had drugs in clinical trials; investors could be confident that an exit, through a trade sale or an IPO, would take place within 4 or 5 years. In the following year Keen and Evans, together with Rea Brothers, a City investment firm, set up the Reabourne Merlin Life Sciences Investment Trust to invest in quoted and unquoted biotech companies, mainly in the UK and other parts of Europe.

Meanwhile the government—still Conservative until Labour's victory in the 1997 election—continued to encourage private investors to invest in entrepreneurial firms. In 1994 the Business Expansion Scheme, which had been misused as a tax-efficient method for investing in property, was replaced by the Enterprise Investment Scheme (EIS); the investment rules were tightened, while the tax advantages for investors in small unquoted firms were increased. The EIS was targeted at wealthy individuals who could invest directly in companies of their choice. For other investors the government introduced Venture Capital Trusts, which were set up and run by fund managers, listed on the stock market, and invested in a portfolio of unquoted firms. By investing in VCTs rather than through the EIS, investors were exposed to less risk, but still enjoyed tax advantages. Income from their investments was subject to a lower rate of tax, and any capital gains were exempt from tax. These schemes proved to be a useful stimulus for investors and entrepreneurs, attracting capital for start-ups across the economy as a whole, including early-stage biotech firms.

The Universities Join In

With the greater availability of venture capital and growing support from institutional and private investors, the UK was beginning to create a financing environment for biotech that could bear comparison with the US. What was still lacking was the partnership between university scientists, entrepreneurs, and financiers that had proved so productive in the US, especially in the San Francisco Bay Area and in Boston.

In Cambridge the first steps in technology transfer were taken in the early 1970s with the establishment of the Wolfson Cambridge Industrial Unit. Supported by a grant from the Wolfson Foundation, it had the task of promoting cooperation between the engineering department and industry, but its scope was broadened in the early 1980s to act as an industrial liaison office for the university as a whole; the commercialization of intellectual property was added to its remit in 1987.[77]

One of the academics supported by the Wolfson unit was Herman Waldmann in the Department of Pathology. His research, partly funded by the MRC, led at the end of the 1970s to alemtuzumab (Campath), a monoclonal antibody which was used first for cancer patients to improve the outcomes of bone marrow transplantation procedures. The original versions of Campath used mouse and rat derived antibodies, but collaboration with Gregory Winter and his colleagues at the MRC's Laboratory of Molecular Biology produced a humanized version of the drug (less likely to provoke an immunogenic reaction in patients), known as Campath 1H, which was developed as a treatment for chronic lymphocytic leukaemia, and later for multiple sclerosis.

The licensing of Campath was handled by BTG, which tried unsuccessfully to interest Celltech in the drug (this was in the early 1980s, when Celltech had not yet established a clear drug development strategy) and then reached an agreement with Wellcome Foundation. Wellcome spent some £50m on Campath development, focusing in particular on rheumatoid arthritis, but decided in 1994, after disappointing clinical trials, to discontinue the programme. (Wellcome was taken over in the following year by Glaxo.) Three years later BTG licensed the drug to a small US biotech firm, LeukoSite, which had been supporting Waldmann's research. In 2001, in collaboration with another US

---

[77] In 2006 all the university's activities in technology transfer and consultancy were transferred to a new company, Cambridge Enterprise.

firm, ILEX Oncology, LeukoSite obtained FDA approval to market Campath as a treatment for chronic lymphocytic leukaemia. After several changes of ownership Campath passed into the hands of Genzyme, one of the biggest US biotechs, which was itself acquired by Sanofi, the French pharmaceutical group, in 2011. Sanofi relaunched the drug 2 years later under a new name (Lemtrada) and at a higher price, as a treatment for the relapsing/remitting form of multiple sclerosis.[78]

Throughout this period Waldmann (who had moved in 1994 to Oxford as head of the Sir William J. Dunn School of Pathology) was closely involved in Campath's progress, as was BTG. In the year ending March 2015, BTG received £4.9m in royalties from Campath/Lemtrada, which were shared with GlaxoSmithKline (based on the original agreement with Wellcome), the MRC and Cambridge University. Although Lemtrada was competing in a crowded field, at launch it was seen as a potential blockbuster drug, with sales of over $1bn per year.

Licensing to an existing company was one option for the commercialization of university research. Another was to start a new firm, as Waldmann's colleague, Alan Munro, did in 1987 when he founded Cantab Pharmaceuticals. But Cantab was not funded or supported in any other way by the university. It was not until the end of the decade that the university began to involve itself directly in promoting spin-out firms. Together with 3i, it set up the Cambridge Quantum Fund to supply seed funding to promising high-technology start-ups, and several other venture capital firms established offices in Cambridge. At the same time a growing number of biotech firms from outside the city established themselves in the Cambridge area. Cambridge was beginning to develop some of the characteristics of a US-style biotech cluster, although it was far behind the other Cambridge, in Massachusetts, in the number and size of its spin-out firms.[79]

Oxford was slower than Cambridge to see the value of collaboration with industry in biotechnology. In the early 1970s 'the prevailing university culture in Oxford was one of hostility or indifference to high-tech enterprise', and this was reinforced by a negative attitude on the part of the local authorities towards industrial development in and around the city.[80] It was not until

---

[78] The Campath story is told by Lara Marks in 'The life story of a biotechnology drug, Alemtuzumab', http://whatisbiotechnology.org/exhibitions/campath.

[79] Steven Casper and Anastasios Karamanos, 'Commercialising science in Europe: the Cambridge biotechnology cluster', *European Planning Studies*, 11, 7 (2003) pp. 805–22. See also 'The biotechnology cluster in Cambridge', in Segal Quince Wicksteed, *The Cambridge Phenomenon Revisited, Part Two* (Cambridge: Segal Quince Wicksteed, June 2000); Philip Cooke, 'Biotechnology clusters in the UK: lessons from localisation in the commercialisation of science', *Small Business Economics* 17 (2001) pp. 43–59.

[80] *Enterprising Oxford, the Growth of the Oxfordshire High-tech Economy* (Oxford: Oxfordshire Economic Observatory, 2000).

the 1980s that the Oxford Science Park, a joint venture between Magdalen College and Prudential Assurance, was established. It was followed by the development of Milton Park, near Abingdon, 12 miles south of Oxford. This was the brainchild of two entrepreneurs, Ian Laing and Nick Cross, who bought the park in 1984, refurbished it, and made it suitable for small, high-technology businesses.[81] They subsequently sold the park to a property company, and used the proceeds to invest on their own behalf in technology-based companies.

Laing and Cross became leading members of an Oxford-based business angel community, working with academics in creating start-up firms. Their first successful investment was in Oxford Asymmetry, founded in 1992 to exploit technology developed by Steve Davies, a professor in the chemistry department. Oxford Asymmetry was not pursuing its own drug discovery programmes; it was a revenue-producing service firm that supplied technology to pharmaceutical companies. In 2000, when the shares of biotech and biotech-related firms were going through an extraordinary boom, Oxford Asymmetry was sold to Evotec of Germany for £316m. Laing and Cross also provided part of the initial funding for Oxagen, and they went on to support several other biotech firms in the Oxford area.

As noted earlier, the first biotech spin-out from the university, Oxford GlycoSciences, was established in 1988, and it was followed by Oxford Molecular, founded by Graham Richards, head of the chemistry department. This company, which used computational techniques to improve the search for molecules that might have therapeutic value, had difficulty raising finance—as a software company it was seen as vulnerable to the departure of a few key software experts—but it secured support from the venture capital arm of Barings, the merchant bank, and several private investors.[82]

Richards was involved in the creation of the university's technology transfer company, Isis Innovation, as was another Oxford academic, John Bell. A Canadian, Bell was then a senior researcher in the Nuffield Department of Clinical Medicine.[83] He had spent several years in California and had seen how US universities had helped to commercialize advances in biotechnology. He was convinced that Oxford must follow their example, a conviction underlined by observing the fate of what might have become a major Oxford-based biotech company.[84]

---

[81] Douglas Hague and Christine Holmes, *Oxford Entrepreneurs* (Oxford: Said Business School, September 2006) pp. 138–140.

[82] Graham Richards, *Spin-Outs, Creating Businesses from University Intellectual Property* (Petersfield: Harriman House, 2009) ch. 5.

[83] John Bell was elected to the board of Roche in 2001 and appointed to the Regius Chair in Medicine in 2002. Knighted in 2009, he became an adviser to the government on policy towards the life sciences.

[84] Hague and Holmes, *Oxford Entrepreneurs*, p. 97.

Garth Cooper, a post-doctoral researcher from New Zealand, had discovered the hormone amylin and showed how it could be the basis for amylin-replacement therapy in diabetes. The commercial potential was clear, and an American venture capitalist, Howard (Ted) Greene, who had started Hybritech and several other US biotech firms, was ready to support the creation of a new company in Oxford. But the university was unhelpful. As Greene recalled later, apart from two professors, Edwin Southern, who ran the biochemistry labs, and Jack Baldwin, head of chemistry, the university authorities 'were really hostile. . . . We started our first operation there in Oxford, in the labs, and when it came time to really set up our own facilities we couldn't. So we moved over to San Diego and I remember Jack Baldwin got up and made quite a noise in Oxford about how stupid they were to be driving these little ventures out.'[85] The company, Amylin Pharmaceuticals, was founded in San Diego in 1987; in 2012 it was sold to Bristol-Myers Squibb for $5.3bn.

This was an attitude that Bell was determined to change, and by the early 1990s the environment for new business ventures was improving.[86] Other universities were also encouraging their scientists to explore commercial opportunities. In London Imperial College had set up its technology transfer arm, later called Imperial Innovations, in 1986; University College did the same 3 years later with the creation of UCL Ventures. Both universities began to launch spin-outs in the late 1990s and early 2000s. As more firms were set up, old inhibitions about involvement in money-making activities began to fade; a growing number of academic scientists were joining with venture capitalists to start new ventures.[87] American ideas about how to commercialize academic science were taking hold.

Blows to Confidence

The UK biotech sector in the mid-1990s seemed to be gathering momentum. Of the three companies that had been regarded as the front-runners, Porton, British Biotech, and Celltech, only the first had disappointed investors. The anti-herpes drug on which great hopes had been placed proved ineffective in

[85] Howard Greene, interview conducted by Matthew Shindell, 8 October 2008, The San Diego Technology Archive (SDTA), UC San Diego Library, La Jolla, CA, lines 454–64, http://libraries.ucsd.edu/sdta/transcripts/greene-howard_20081008.html. Greene went on to say of Oxford University: 'Today [2008] they have a science park and that attitude has changed completely.'

[86] Bell was later involved in several Oxford start-ups, including PowderJect (1993), Oxagen (1997), and Avidex (1999).

[87] The biggest sources of spin-outs were Oxford, Cambridge, Imperial College, UCL, and KCL, but there were several founders from other institutions, including David Lane from Dundee (Cyclacel), Lindy Durant from Nottingham (Scancell), Mark Ferguson from Manchester University (Renovo), and Paul Workman from the Institute of Cancer Research (Chroma and Piramed).

clinical trials, and the company had failed to make good use of the large amounts of cash which investors had provided. Haydon-Baillie resigned as chairman in 1992, and the new management scaled back its expenditure and its ambitions. In 1994 Porton was bought by Ipsen of France for £65.5m.[88]

Meanwhile, British Biotech was riding high. In 1995 investors were encouraged by news of positive Phase II clinical data for its pancreatitis treatment; it also made a partnership deal with Glaxo Wellcome for the development of a new asthma drug.[89] McCullagh's strategy was to retain the rights for potential new hospital medicines but to form partnerships with pharmaceutical companies to fund the development of drugs sold to general practitioners. Specialist drugs needed a smaller sales force, and for these he planned to build his own marketing organization. He was determined to obtain full value from what was expected to be the company's most profitable drug, marimastat, a treatment for cancer. Shareholders were enthusiastic, lifting the British Biotech share price to a peak of 327p in May 1996; the value of the company briefly climbed above £2bn.

Celltech shares also performed well in response to good news from clinical trials and new partnership deals. In July 1994 the shares rose by more than 10 per cent on news of a collaborative agreement with Merck on an asthma treatment that could generate revenues of more than $100m a year.[90] A few months later Celltech reported what it described as a breakthrough in Phase II clinical trials of an anti-arthritis drug which was partnered with Bayer.

Yet, as US experience had shown, shares in drug discovery firms could go down as fast as they went up. An ominous sign of fragility came early in 1996 when Merck announced that it planned to discontinue its partnership to work on Celltech's asthma drug. Celltech was given 45 minutes' notice of Merck's decision and when it was announced the share price fell by 24 per cent.

This episode highlighted the risks involved in partnerships with Big Pharma, and strengthened Peter Fellner's view that Celltech's financial position needed to be made more resilient. That was the rationale for the decision in June 1996 to sell its manufacturing arm, Celltech Biologics, to Lonza, a Swiss company, for £50m. Although the manufacturing side was the company's main source of revenue, the deal was welcomed by investors because it reduced the risk that the company would ask the stock market for more capital. Fellner said the deal would improve the company's risk–reward balance and allow it to concentrate on drug development.[91]

---

[88] *Financial Times*, 27 October 1994. Porton's most valuable asset was Dysport, a botulinum toxin product used in a range of therapeutic and aesthetic applications.
[89] Ernst & Young, *European Biotech 96, Volatility and value*, March 1996.
[90] *Financial Times*, 22 July 1994.   [91] *Financial Times*, 13 June 1996.

Another blow came in 1997 when the drug for Crohn's disease which Celltech had been developing with Bayer failed in its final trials. As the *Financial Times* reminded its readers, success in biotechnology could be a long time coming. 'Celltech has been in business for 17 years and has still not brought a drug to the market.'[92] The contrast with Genentech and Amgen, both of which had produced drugs within less than 10 years of foundation, was stark. (The comparison was misleading, since the two US companies had started by using recombinant DNA-based manufacturing methods to produce known protein therapies, whereas Celltech was trying to produce novel drugs with novel monoclonal antibody technologies.) Cellech's most advanced drug candidate, an antibody-based treatment for leukaemia, gemtuzumab (Mylotarg), partnered with Lederle in the US, was approved by the FDA in 2000.

More serious was a dramatic change of fortune at British Biotech. In 1996, flush with cash after the capital issues that had followed its flotation, this firm seemed on course with its two lead drugs. But in 1997 the European Medicines Agency failed to approve the pancreatitis drug, Zacutex, requesting more data. At the same time, when British Biotech was raising further funds in London to support the development of the cancer drug, marimastat, the company made it clear that results would not be available until 1999. The shares began to weaken and by early 1998 they were down from their peak of over 300p to 91p.

To make matters worse, there was dissension among the senior executives, stemming from disagreements over strategy as well as a dispute over whether management should be allowed to sell shares. The first public sign of board tensions was the resignation in late 1996 of John Gordon, research director, followed in February 1997 by that of the finance director, James Noble; Peter Lewis, director of research and development, left a few months later.[93]

Meanwhile the head of clinical research, Andrew Millar, had become uneasy about the tone of the company's press statements about the progress of clinical trials.[94] Early in 1998 he revealed his concerns to an analyst at Goldman Sachs and then to two representatives of Perpetual, the unit trust group which held 9.5 per cent of British Biotech.[95] Millar also met with Mercury

---

[92] *Financial Times*, 24 May 1997.

[93] Peter Lewis told MPs later that he had been concerned that the company's commercial strategy was too ambitious; he thought British Biotech should have been more willing to license its drugs to established companies rather than commercializing them itself. Memorandum submitted by Dr Peter Lewis to House of Commons Select Committee on Science and Technology, 13 July 1998.

[94] Jonathan Guthrie, 'The changing fortunes of British Biotech', *Financial Times*, 2 May 1998.

[95] The two Perpetual managers were Margaret Roddan and Neil Woodford. The latter was to become one of the most influential investors in UK biotech, first with Perpetual and then with Invesco, which merged with Perpetual in 2000. He broke away from Invesco in 2014 to set up his own investment firm (see Chapter 5).

Asset Management, which had an 11.5 per cent stake, and with Dresdner Kleinwort Benson, the company's financial advisers.

The board commissioned an investigation by an independent law firm, which concluded that there had been no justification for Millar's allegations against McCullagh's management. However, Millar was dismissed in April for discussing confidential information with third parties.[96] He repeated his concerns in a letter to the *Financial Times*, an edited version of which was published on 29 April 1998.

Investors, disturbed by Millar's disclosures, lost confidence in the management; by the spring of 1998 the share price had fallen to just over 50p. In May McCullagh announced his intention to step down as chief executive.[97] He was replaced in September by Elliott Goldstein, a senior manager at SmithKlineBeecham.[98]

The British Biotech affair was a spectacular fall from grace which was to sour attitudes to the UK biotech sector for many years to come.[99] Another high-flying company which fell to earth was Scotia, which had been floated in 1993.[100] The failure to secure regulatory approval for Tarabetic, a treatment for diabetes, led to a sharp fall in the share price. By the end of 1997 the company was running out of cash and David Horrobin, the founder, resigned. His successor, Robert Dow, came from Roche, as did the new director of drug development, Chris Blackwell. (Blackwell later took over from Dow as chief executive.) Scotia was a case, not unusual in biotech, where the scientist-founder had stayed on too long and had pursued too many disparate projects. Investors were looking for new leaders who had managerial and commercial experience as well as scientific skills.[101]

---

[96] These events are described in the fifth report of the House of Commons Select Committee on Science and Technology, published on 17 August 1998.

[97] McCullagh was later appointed chief executive of Santaris Pharma, a Danish firm which was bought by Roche, and chairman of Xention, a British biotech firm.

[98] British Biotech halted work on marimastat in 2001. In 2002, data from a separate clinical trial of marimastat was published, showing that the drug significantly reduced mortality in advanced gastric cancer. S. R. Bramhall et al., 'Marimastat as maintenance therapy for patients with advanced gastric cancer: a randomised trial', *British Journal of Cancer* 86 (2002) pp. 1864–70. Marimastat was a small molecule drug designed to inhibit an enzyme called MMP9 (matrix metalloproteinase). More recently Gilead in the US has developed a monoclonal antibody which is a selective MMP9 inhibitor; it is showing promise in ulcerative colitis and solid tumours.

[99] As late as July 2014 the CEO of a leading US biotech company, interviewed by the authors in Boston, referred to the British Biotech case as an important part of the reason for the disappointing performance of the UK biotech sector.

[100] Scotia had been founded by David Horrobin in the 1970s as Efamol Holdings to make evening primrose oil as a dietary supplement. Horrobin used the profits from this business to develop a pipeline of drugs, mainly based on fatty acids. The company was renamed Scotia and came to be classified as a biotech firm.

[101] *BioCentury*, 15 December 1997.

These and other setbacks unnerved investors.[102] Too often, it seemed, ambition had run ahead of what was feasible, and at the peak of the boom in 1996 valuations had lost touch with reality. As one analyst remarked, 'there weren't that many people in London who knew how to value these things properly. We are all on a steep learning curve—investors, analysts, financiers and the companies themselves.'[103] Another participant, reflecting on these events some years later, said: 'The over-optimism of biotechs in the UK coupled with the naiveté of investors was a disastrous toxic mix.'[104]

One consequence was a retreat on the part of investing institutions. The most striking withdrawal was that of Rothschild, whose two investment trusts, Biotechnology Investments Limited and International Biotechnology Trust, had been performing poorly.[105] In March 1999 Rothschild announced that it was withdrawing from biotech investment in the UK. BIL was taken over by 3i, and the management of IBT was transferred to Schroders.

This was part of a general shift within the UK venture capital community away from high-risk, early-stage businesses (in information technology as well as biotech) to management buy-outs, which offered a more predictable and quicker return. Those that remained committed to biotech preferred firms that had drugs in late-stage trials and were likely to generate revenue in the near term.

First Steps Towards Consolidation

Peter Fellner at Celltech had concluded after the failed collaborations with Bayer and Merck that Celltech needed to be less dependent on partnerships, which, even if they worked well, involved giving away a large proportion of the profits from any successful programme. He saw acquisitions as a means of creating a more resilient and more diversified business, and reducing the binary risk in biotechnology—over-dependence on a single drug programme which, if it failed, would put the whole company in jeopardy.

His first target was Chiroscience, which since its flotation in 1993 had brought its local anaesthetic, Chirocaine, to the market; it also had a stake in

---

[102] Cantab, the Cambridge-based firm, ran into problems at the end of the 1990s when its lead vaccine for the treatment of genital warts failed in Phase II trials. Cantab was bought by Xenova in 2001; the merged company, after several disappointments in clinical trials, was sold to Celtic Pharma, a private equity group, in 2005. Another casualty was Cortecs, a drug delivery firm with Australian origins which listed in London in 1994 and reached a market value of over £600m in 1996. After failures in clinical trials investors lost confidence in the management and the market value fell to £24m at the end of 1998. The company was renamed Provalis and continued in business until 2006, when it was liquidated.

[103] *Financial Times*, 16 February 1999.

[104] Private communication.   [105] *Financial Times*, 27 March 1999.

the emerging field of genomics, through the acquisition of Darwin Molecular in Seattle. However, the attraction for Fellner lay not so much in Chiroscience's portfolio as in its cash. The takeover, agreed in July 1999, was a share-based transaction which valued Chiroscience at £331m, giving investors in that company a healthy return. John Jackson, Celltech's chairman, said the merger marked a watershed in the history of European biotechnology. 'What we have done may have a catalytic effect', he said; 'up till now there has been a sense of paralysis in the industry'.[106]

Fellner's next and much bigger acquisition was that of Medeva, which had been built up by Ian Gowrie-Smith and his successors mainly through in-licensing and acquisitions. This deal, which followed an earlier approach to Medeva from Shire, went through in 1999 and valued the company at £563m. Medeva had some mature but profitable over-the-counter businesses in the US, as well as several late-stage or recently approved drugs, including Metadate, a treatment for attention deficit hyperactivity disorder (ADHD). It was generating about £100m in operating cash flow, which would help to finance Celltech's R & D programmes.

Rolf Stahel at Shire also believed in acquisitions, but his targets were in the US rather than the UK. Soon after going public in 1996 Shire made two US acquisitions, Pharmavene, a drug delivery company, and Richwood Pharmaceutical, whose principal product was Adderall, an ADHD drug; much of the growth in Shire's sales and profits over the next few years came from Adderall. Soon after these deals had gone through Shire listed on the NASDAQ, making its shares more accessible to US investors.

The next step, in 1999, was the $900m takeover of a larger US company, Roberts Pharmaceutical, which gave Shire a position in oncology and gastro-intestinal disorders. As a result of these deals some 70 per cent of the company's shares were held in the US, but Shire retained its London listing and its development laboratory in Basingstoke. Stahel was greatly admired by investors, and Shire's high share price gave him the currency with which to finance further acquisitions.

These deals lifted Celltech and Shire well above the crowd of small drug discovery firms that constituted the bulk of the UK biotech sector; both were valued in the stock market at well over £1bn at the end of the 1990s. Whether more consolidation among the smaller firms was desirable was a much discussed question at that time. Although several amalgamations took place, a later study concluded that mergers among small UK biotechs did not generally lead to stronger pipelines or make the merged company more attractive to potential partners.[107]

---

[106] *Financial Times*, 16 June 1999.

[107] Vanessa Maybeck and William Bains, 'Small company mergers—good for whom?' *Nature Biotechnology* 24 (2006), pp. 1343–8.

For drug discovery firms based on innovative science, the principal challenge was not to get bigger by acquisition but to validate their technology through partnerships and licensing agreements with pharmaceutical companies, and to use the revenue generated by these deals to finance their drug development programmes. That was what Genentech had done in the US, and it was what one of the most promising UK firms, Cambridge Antibody Technology, had been trying to do since its foundation in 1989. By 1997 CAT had made sufficient progress—it had partnerships with several leading companies, including Genentech, BASF, Pfizer, and Eli Lilly—to permit a stock market flotation; its subsequent evolution is described in the next chapter.

UK Biotech at the End of the 1990s

At the end of the 1990s the UK biotech sector, as one chief executive put it, was 'still in its teenage years'.[108] Since the foundation of Celltech in 1980 the UK had moved part of the way towards building the institutions which had underpinned the growth of biotech in the US. A sizeable venture capital industry had been established; the stock market was accessible to early-stage firms; and the links between universities and business were improving. What remained a matter of concern was the absence of big successes—firms that had shown themselves capable of bringing drugs to the market, making profits, and retaining the support of investors as they continued to grow. The only major exception was Shire, and Shire's success was not associated with drugs originating from the UK; the company was increasingly focused on the US.

There was no sign that the gap between the UK and the US in biotech was narrowing. Even in Europe, although the UK biotech sector was larger than that of any other country, there were worries about how long that lead could be maintained. Germany, in particular, had set in train an ambitious programme aimed at creating strong biotech clusters comparable to those in Boston and San Francisco; the stated aim was to overtake the UK in biotech (see Chapter 6). At the same time the Frankfurt Stock Exchange had taken steps to improve the flow of equity capital into early-stage firms by setting up a separate market, the Neuer Markt; as with AIM in London, the new market's listing rules were less demanding than on the main market, allowing loss-making firms to go public.

The competitive threat from Germany was considered serious enough for Lord Sainsbury, Science Minister in the new Labour government, to ask a

---

[108] Robert Mansfield, chief executive of Vanguard Medica, quoted in *Financial Times*, 26 October 1998.

group of industrialists and academics in 1999 to examine the scope for strengthening biotech clusters in the UK; the group included Chris Evans, founder of Chiroscience, and Professor Mark Ferguson of Manchester University, who had recently founded Renovo, a biotech firm that specialized in anti-scarring treatments.[109]

Although Lord Sainsbury's report led to no major policy changes, his intervention was a sign of the government's anxiety over the future of a sector which was still seen as one of the country's rising stars but seemed not to be fulfilling its early promise.[110] That anxiety was to intensify over the next few years as the performance of the sector deteriorated.

---

[109] *Financial Times*, 16 April 1999.
[110] *Biotechnology clusters*, Report of a team led by Lord Sainsbury, Minister for Science, Department of Trade and Industry (London: HMSO, August 1999).

4

Investors Retreat

The BIGT's vision is that by 2015 the UK will have secured its position as a global leader in bioscience. To secure this position the UK must create a diverse, self-sustaining bioscience sector with a core of large, profitable, world class companies, second in size and achievement only to the US. Report by Bioscience Innovation and Growth Team to the UK government, 2003[1]

The first decade of the new millennium saw much of the optimism that had surrounded the birth of UK biotech fade away. Several of the leading companies lost their independence through acquisitions by Big Pharma or by other biotech firms. Although new firms continued to be created, their owners generally expected to exit their investments through a trade sale rather than an IPO. This partly reflected the diminishing interest in UK biotech on the part of institutional investors. With venture capitalists also retreating from the sector, the financing environment was much less favourable than it had been in the 1990s, prompting some firms to move to the US, where the pool of investors committed to biotech was much larger. The quality of British science remained a valuable asset, but there was concern among policy makers about the dearth of large or even medium-sized British-owned biotech firms capable of exploiting it.

A Short-lived Boom

The new millennium began with a short-lived boom in biotech shares on both sides of the Atlantic as investors who had piled into internet-related businesses in the late 1990s shifted their attention to what they saw as the 'next

---

[1] *Bioscience 2015, Improving National Health, Increasing National Wealth*, A Report to the UK Government by the Bioscience Innovation and Growth Team (Department of Trade and Industry, November 2003).

big thing'. The imminent completion of the Human Genome Project was thought to herald a wave of innovation in genomics-based drugs; in the US, as discussed in Chapter 2, the shares of firms that had a stake in genomics were pushed to extraordinary heights. In the UK there were only a few small firms that had some claim to being part of the genomics wave, but investor enthusiasm was channelled to other drug discovery firms that were using novel technologies, including proteomics and monoclonal antibodies.[2]

In the UK two of the biggest stock market gainers were Cambridge Antibody Technology (CAT) and Oxford GlycoSciences (OGS), both of which took advantage of the boom to raise more money from investors. Ten biotech or biotech-related firms floated on the stock market in 2000, five on the LSE's main market and five on AIM. There was also an increase in venture capital financing. In one of the largest private financing rounds Oxagen raised £30m from a group of investors that included SV Life Sciences, Advent, and Abingworth. Oxagen was a platform company, a business model that was popular with investors at that time, using advances in genetics as a means of identifying the link between genetic variations and disease, and selling that information to pharmaceutical companies.

Towards the end of 2000 shares began to weaken and the slide continued in 2001 and 2002. Several of the companies which had floated during the boom saw their share price fall sharply. An extreme case was that of ReNeuron, a company that had been spun out of King's College London, with the backing of Chris Evans's Merlin Fund, to develop stem cell technology for therapeutics. The shares were floated at 195p in 2000 and reached a peak of 225p in the following year. But the company was hit by technical problems—the cell lines it was developing, derived from human foetal cells, became unstable over time—forcing a delay in clinical trials. Several attempted partnerships did not materialize, cash reserves were declining, and by February 2003 the shares were worth less than 5p. In April of that year ReNeuron was taken private through a management buy-out, with Merlin continuing as the largest shareholder.

To add to the gloom there were several casualties among firms that had floated in the 1990s. Scotia, under new management following the removal of the founder-chairman in 1997, had pinned its hopes on temoporfin (Foscan), a photodynamic therapy targeted at cancer.[3] When Foscan was rejected by the Food and Drug Administration in the US at the end of 2000 and by the

---

[2] P. Martin, M. M. Hopkins, P. Nightingale, and A. Kraft, 'On a critical path: genomics, the crisis of pharmaceutical productivity and the search for sustainability', in *The Handbook of Genetics and Society* (London: Routledge 2009) edited by P. Atkinson, P. Glasner, and M. Lock, pp.145–62.

[3] Photodynamic drugs are inactive until exposed to a specific wavelength of light, which activates them at a particular location, allowing therapeutic effects to be focused at particular sites in the body.

European regulators a month later, Scotia went into administration. Although there was some prospect of reversing the European decision, as indeed happened later in 2001, Scotia could not raise enough money to stay afloat.[4]

A more spectacular collapse was that of Bioglan Pharma, a skin care specialist. Founded in 1985 and floated in 1998, it achieved a market capitalization of over £800m in early 2000. A series of setbacks, including the failure to complete a planned acquisition of Bristol-Myers Squibb's dermatological unit, undermined the confidence of investors. At the end of 2001 Bioglan shares were 98 per cent lower than at the start of the year; it went out of business in February 2002.

For listed firms any piece of bad news was likely to precipitate a decline in the share price, making it impossible to raise new funds without heavily diluting the value of shares held by existing investors. When PPL Therapeutics joined the stock market in 1996, its lead drug, a treatment for cystic fibrosis and emphysema, was about to enter clinical trials. The project relied on the production of a recombinant therapeutic protein, alpha-1-antitrypsin (AAT), harvested from the milk of transgenic sheep. This proved more difficult than expected, but when Bayer came in as a partner in 2000 there were hopes that the drug could reach the market within a few years and achieve annual sales of at least £250m; the German pharmaceutical company was the principal producer of AAT extracted from human plasma, which was in short supply. The trials on AAT did not yield the hoped-for results and when Bayer pulled out in 2003 PPL was left with not much cash and limited prospects. An attempt to restructure the company around a non-therapeutics business failed, and PPL was closed down.[5]

Capital Availability

'If any good comes out of the current downturn', Ernst & Young wrote in its 2003 European biotech report, 'it will hopefully be an increased focus on revenues, operational cash flow and profits. The days are gone, for the time being at least, when a biotechnology company with interesting technology could expect to set out with a business plan that showed ten years until it received revenues and fifteen or twenty years to profitability.'[6]

UK venture capital firms saw management buy-outs as more lucrative and less risky than investing in loss-making high-technology businesses, and this

---

[4] The Foscan business was sold by the administrators to Singapore Technologies for £70m and later acquired by a German company, biolitec pharma, which continues to market it.

[5] PPL's patent portfolio, covering among other things its recombinant protein technology, was sold to Pharming of the Netherlands.

[6] Ernst & Young, *The European Biotechnology Report* (2003).

applied as much to information technology as to biotech. The few that remained committed to biotech shifted more of their investment from early-stage firms to ones that had late-stage products near to commercialization, and to medical devices and diagnostics, where the path to profitability was more predictable. With the low valuations of biotech companies on public markets, some of them took the opportunity to invest in the sector through the use of PIPEs—private investment in public equity—a mechanism that was widely used in the US whereby venture capital firms bought shares in firms that were already listed on the stock market.

Yet the old model of venture-backed early-stage firms was not dead. In 2003 two of the most successful VCs, Abingworth and SV Life Sciences, had little difficulty in raising new funds ($350m in the first case, $402m in the second). Kate Bingham, general partner at SV Life Sciences,[7] commented that biotech firms could no longer float their shares on the stock market as early as they had done in the 1990s and so were more dependent on venture capital to keep their programmes going. She would be encouraging more companies to merge so as to build critical mass.[8]

Some of the major stockbrokers who had brought clients to the stock market in the 1990s did not return, reflecting the fact that most institutional investors, having had their fingers burned in Porton, British Biotech, and others, no longer had much appetite for UK biotech. Yet there was one notable exception. Perpetual, the unit trust group which had been an investor in British Biotech and other first-generation firms, was acquired in 2000 by Amvescap, an American fund management group that was later renamed Invesco; the UK business was called Invesco Perpetual. In 2002 Amvescap took advantage of the depressed share prices to build up stakes in a range of UK biotech firms including OGS, CAT, Celltech, and PowderJect.[9] The assumption, which proved to be correct, was that many of these firms would either recover or be acquired at a higher price.

An important influence on Invesco's policy was Neil Woodford, the ex-Perpetual fund manager who had been an investor in British Biotech at the time of that company's crisis in 1998. Woodford was to become one of the UK's most successful fund managers, and a key supporter of UK biotech (as discussed in later chapters). In the 1990s he ran two large funds, the bulk of which were invested in well-established, dividend-paying companies. But he allocated a small part of his portfolio to high-risk, science-based firms, mainly in biotech, which he thought had the potential to become big winners. He was

[7] Kate Bingham had studied biochemistry in Oxford before starting work as a management consultant. After taking an MBA at Harvard she worked in business development at Vertex, the US biotech company, before joining Schroders in 1991.

[8] *Financial Times*, 4 June 2003.

[9] Patrick Jenkins, 'Opportunities for investors in battered sector', *Financial Times*, 1 July 2002.

particularly interested in promising firms spun out from universities, and he supported two university-linked organizations which specialized in the commercial exploitation of academic discoveries.

One of them sprang from an unusual collaboration between Graham Richards, professor of chemistry at Oxford, and David Norwood, head of a City stockbroking firm, Beeson Gregory.[10] Under an agreement signed in 2000, Beeson Gregory invested £20m in a new laboratory for the Department of Chemistry, in return for the right to a 50 per cent share in any spin-outs from the department for the next 15 years. Beeson Gregory merged in 2003 with Evolution Securities, and the new group established a subsidiary, IP2IPO ('intellectual property to initial public offering'), which managed the partnership with Oxford and formed similar arrangements with other universities. Run by Norwood and later renamed IP Group, this company was floated on AIM in 2003 and switched to the main market in 2006. Invesco was not an investor at the start but acquired a substantial stake in 2011.[11]

Woodford also supported Imperial Innovations, which had evolved from the technology transfer office of Imperial College to become a separate company, majority-owned by the college but with outside shareholders. By 2006 it had a portfolio of ninety-six licensing agreements and fifty-eight spin-out companies in a range of sectors.[12] In that year, in order to access larger amounts of capital and to give its existing investors a value for their shares, Imperial Innovations listed on AIM. The college retained a controlling interest, but its shareholding was later reduced to below 50 per cent as Invesco and other institutions bought more shares.

IP Group and Imperial Innovations, unlike conventional venture capital firms, were not obliged to exit their investments within a specified time period, making them potentially longer term investors. On the other hand, they were commercial organizations whose investors expected a return. In the 1990s that return would have been achieved through floating the most advanced of their portfolio firms on the stock market. After the collapse of the 2000 boom that option was no longer available, at least as far as the LSE's main market was concerned; there were no IPOs between 2001 and 2003, four between 2004 and 2006, then none until 2014 (Table 4.1).

For firms seeking smaller amounts of capital a flotation on AIM was still possible. In 2001 GW Pharmaceuticals, which was developing cannabis-based drugs for multiple sclerosis and other diseases, raised £25m on AIM. It continued to trade on AIM long enough to bring its lead drug, Sativex (nabiximols),

---

[10] Graham Richards, *Spin-outs* (Petersfield: Harriman House 2009) pp. 123–4. David Norwood, a chess grandmaster, had previously run an information technology consulting firm, Index IT.

[11] In 2015 Invesco held a 25.7 per cent share of the IP Group.

[12] *Financial Times*, 21 July 2006.

Table 4.1. Biotech IPOs on the LSE main market, 2000–2014

Company	Amount raised (£m)	Market capitalization at IPO (£m)	Date
Pharmagene	40.7	142.0	2000
Weston Medical	52.4	210.1	2000
Profile Therapeutics	26.0	83.5	2000
Ark Therapeutics	55.0	118.0	2004
Ardana	21.0	71.1	2005
ProStrakan	40.0	186.8	2005
Renovo	50.0	154.1	2006
Circassia	200.0	583.0	2014

Source: London Stock Exchange.

closer to regulatory approval. (GW's recent history is discussed in the next chapter.)

GW was followed onto AIM by Neutec Pharma, a 3i-backed spin-out from the University of Manchester, which raised £10m in 2002. Neutec was developing a pipeline of anti-infective drugs based on monoclonal antibodies; its technology looked sufficiently promising to attract a takeover offer from Novartis in 2006, which at £305m came at a 109 per cent premium to the stock market share price.[13]

That some investors were still willing to support high-risk drug discovery firms on AIM was encouraging, but the absence of IPOs on the main market, coupled with the decline in venture capital support for early-stage firms, threatened to stunt the growth of a sector that had been seen as one of British industry's rising stars. This was a situation that the Labour government, which had taken a close interest in biotech since it entered office in 1997, viewed with increasing concern.

The Government Intervenes

At the start of 2003 the government, together with the industry's trade body, the BioIndustry Association, set up the Bioscience Innovation and Growth Team, with a mandate to work out a 'strategic approach' to the future of the industry. The chairman was Sir David Cooksey, head of Advent Venture Partners.[14] The steering group was made up of senior executives in biotech firms, academics, and government officials.

---

[13] Soon after the acquisition Neutec's two leading projects failed: one in clinical trials and the other because of regulatory concerns. *The Independent*, 22 April 2007.

[14] David Cooksey was knighted in 1993. In addition to his role in biotech, he held several public sector appointments, including chairmanship of the Audit Commission from 1986 to 1995.

The team's report, published at the end of 2003, warned that the sector faced an immediate and severe funding crisis.[15] Over a third of publicly quoted UK bioscience companies had less than 2 years of cash remaining and, the report noted, 'it is difficult to see how some of them will raise the money they require to continue'. The medium- to long-term challenge was 'to provide enough funding to create a critical mass of self-sustaining profitable companies with significant ownership of products launched in major markets'.

The report made several recommendations on finance, not all of which were within the government's power to deliver. These included one of the industry's long-standing requests, that the London Stock Exchange should amend the pre-emption guidelines to permit UK-listed companies to issue up to 20 per cent of their share capital on a non-pre-emptive basis—that is, without first offering them to existing shareholders—in any 3-year period. Under the existing rules the limit was set at 5 per cent, which, the industry argued, made raising new capital cumbersome and expensive, as well as making share-based acquisitions of other firms more difficult. The government's response was to commission an inquiry from Paul Myners, a leading institutional investor. His report recommended changes to make the system more flexible, but did not question the right of investors to limit the extent of any dilution that might arise when companies raised new capital.[16] Pre-emption rights continued to be an obstacle for capital-hungry firms when trying to raise new funds.[17]

The BIGT report called for changes in the R&D tax credit scheme, which had been introduced by the Labour government in 2000 as part of a wider set of measures aimed at stimulating investment in research, and for additional financial support to bridge the gap between idea generation and commercial financing. The government rejected some of these proposals, either because they were incompatible with UK tax policy or because they contravened EU state aid rules.[18] In a comment that had an almost Thatcherite ring, the government said it did not support the picking of industrial or sectoral winners for specific, unconditional financial support; 'rather it seeks to identify market failures in order to address the causative factors across the board'.

Central to the Cooksey report was its vision of the future of the industry, which was to create 'a diverse, self-sustaining bioscience sector, with a core of large, profitable, world class companies'. The success of the industry in the near and medium term, according to the report, depended on a handful of

[15] *Bioscience 2015*, A report to the government by the Bioscience Innovation and Growth Team, November 2003.

[16] *A study by Paul Myners into the impact of shareholders' pre-emption rights on a public company's ability to raise new capital*, Department of Trade and Industry, February 2005.

[17] Andrew Wood, 'Direct costs of share issues on public markets—reviving the debate over non-pre-emptive share issues', BioIndustry Association, 12 December 2012.

[18] *The Government response to the Bioscience Innovation and Growth Team*, Department of Trade and Industry (London: HMSO, November 2004).

companies achieving sustainable profitability and thus providing beacons both to the investor community and to the industry itself.

Looking for Beacons

Where were the beacons to come from? Since the British Biotech debacle, Celltech seemed the most likely candidate. After the purchase of Chiroscience and Medeva and the sale of its manufacturing business to Lonza, Celltech was now in a more comfortable financial position. The priority was to push on with its drug development programmes while at the same time keeping an eye out for acquisitions that might strengthen its portfolio. So far only one of its drugs, Mylotarg, a monoclonal antibody treatment for acute myeloid leukaemia which was partnered with Lederle in the US, had been approved by the FDA; it was expected to generate only modest royalties.[19]

High hopes were placed on an anti-TNF biologic drug targeted at rheumatoid arthritis and Crohn's disease, CDP870.[20] For Celltech to undertake the full development of this drug on its own was deemed to be beyond the company's financial capacity, and it therefore sought to negotiate a partnership with Big Pharma. Several companies were interested, and the best offer came from Pharmacia, a Swedish–American pharmaceutical group.

When the Pharmacia agreement was signed in 2001, it was seen as a landmark deal for Celltech. It involved a large up-front payment, sizeable milestones, and a substantial share in profits; the drug was expected to achieve annual sales of over $1bn. The link with Pharmacia was strengthened at the end of 2002 when Göran Ando, who had been that company's head of research, was named as the successor to Fellner as Celltech's chief executive; Fellner became non-executive chairman, succeeding John Jackson.

Early in 2003, shortly before these changes took effect, an opportunity arose for another acquisition. Oxford GlycoSciences (OGS) had come out of the 2000 boom in an apparently strong position. It had large cash reserves and a 'hybrid' business model which appealed to investors.[21] It had both a pipeline

---

[19] Lederle was part of American Cyanamid until 1994, when it was acquired by American Home Products. In 2002 American Home Products changed its name to Wyeth; it was acquired by Pfizer in 2009. In 2010 Mylotarg was withdrawn from sale in the US after FDA concerns about its safety profile.

[20] Anti-TNF drugs block the action of a protein called tumour necrosis factor, which, if overproduced in the body, causes inflammation to the bones and tissue. Celltech had previously developed an anti-TNF drug, partnered with Bayer, for the treatment of septic shock (CDP571), but this had been abandoned after disappointing clinical trials. CDP870 (certolizumab pegol) was a different modality of higher affinity for the target molecule.

[21] Business models in biotech were commonly divided into three categories: the product business model, through which the company sought to generate value by developing its own drugs and either taking them through to commercialization or licensing them out to pharmaceutical companies; the

of drugs under development and a proteomics platform which was expected to be a valuable tool for identifying disease targets.[22] It also had collaborations with several Big Pharma companies. In the short term, however, the company's hopes depended crucially on its lead drug, Zavesca, an oral small molecule drug for Gaucher's disease that offered an alternative to the enzyme replacement therapy developed by Genzyme in the US.[23]

During 2002 OGS investors' confidence was undermined by reports from clinical trials that Zavesca caused serious side-effects in some patients and then by the FDA's decision, in June of that year, not to approve the drug without further evidence. Although Zavesca was later approved by European regulators and subsequently by the FDA, the share price fell sharply in response to the initial FDA ruling; the market value of the company was only slightly higher than its cash.

The chief executive, David Ebsworth (recruited from Bayer to replace Michael Kranda, who had returned to the US), was under pressure to find other late-stage drugs through acquisitions. While the search for new opportunities was under way OGS was approached by CAT with a proposal that the two companies should merge. The proposal was welcomed by commentators as a step towards consolidation in a fragmented industry, as well as creating a European biotech with a cash balance comparable to that of the larger US firms.[24]

The OGS board agreed to the terms offered by CAT, but the merger was to take the form of an exchange of shares, and when the CAT share price fell as a result of a dispute over royalties with Abbott of the US, one of its most important licensees, the proposal had to be abandoned. Celltech stepped in with an all-cash counter-offer, and with the support of the two principal shareholders in OGS, Invesco and Fidelity, carried off the prize.[25]

The OGS takeover was an opportunistic move on Celltech's part, motivated mainly by OGS's large cash pile. The value of Zavesca had been reduced by Ebsworth's decision to sell the European marketing rights to Actelion of Switzerland in return for a royalty; OGS had other drug candidates, principally for cancer, but these were at the pre-clinical stage. The proteomics platform

platform or tools model, through which the company generated revenue by selling technology, tools, or services to other companies; and the hybrid business model which generally included a platform technology capable of generating a pipeline of products, to be used both for licensing to other companies and for in-house drug development. Jane Fisken and Jan Rutherford, 'Business models and investment trends in the biotechnology industry in Europe', *Journal of Commercial Biotechnology* 8, 3 (2002) pp. 191–9.

[22] Proteomics involves the large-scale analysis of the structure and functions of proteins.

[23] Genzyme's product was a recombinant protein drug delivered in hospital via infusion.

[24] *BioCentury*, 27 January 2003.

[25] These two institutions were also shareholders in CAT and Celltech. *Financial Times*, 27 February 2003.

was potentially valuable, but unlikely to generate significant revenue in the short term.[26] However, the merger created a bigger company and consolidated Celltech's status as the flagship of the UK biotech sector.

Whether it could sustain that position depended on its ability to bring drugs to the market. This was Ando's first priority as he took on the role of chief executive, although he was also open to further acquisitions. As he told an interviewer, 'I am looking at a dual pathway of both driving value through our own pipeline and in parallel to this through consolidation . . . this is the best way for us to create a European bellwether.' He saw US firms such as Genentech and Gilead as role models.[27]

Much hinged on the success of CDP870. More safety and toxicology studies were needed, and Pharmacia invested heavily in the analytical and clinical development of the drug. In April 2003, while this work was under way, Pharmacia was acquired by Pfizer, one of the largest US pharmaceutical companies. Although the new owners continued to support CDP870 they insisted on revising the terms which Pharmacia had agreed with Celltech. After much internal debate Fellner and Ando rejected Pfizer's proposed terms and brought the partnership to an end. The board then decided that, since it was not feasible to raise from shareholders the capital needed to complete the clinical trials of CDP870 and to take it through to registration and marketing, a new partner had to be found.

Over the next few months Celltech received approaches from more than a dozen companies which were interested in the drug. Celltech had other assets, including a potential treatment for osteoporosis (partnered with Amgen) which was in early clinical trials, as well as the over-the-counter medicines that it had acquired with Medeva. It was possible that a prospective partner might wish to buy the whole company, and that is what happened in May 2004 when the Belgian pharmaceutical group, UCB, made an agreed takeover offer that valued Celltech at £1.5bn.

The *Financial Times* described the sale to UCB as marking 'another unimpressive chapter in the history of the British biotechnology industry'.[28] Others argued that the deal reflected a realistic assessment by the Celltech board of the need for scale and for sufficient financial resources to fund future projects; Celltech, according to this view, had delivered value for its investors without exposing them to the risk of product failures.[29]

It was true that investors who had bought Celltech shares when they were at their low point, including Invesco, made a good profit from the sale to

---

[26] The proteomics unit was closed down in 2004.   [27] *BioCentury*, 4 August 2003.
[28] Lex Column, *Financial Times*, 19 May 2004.
[29] This was the view of Rolf Stahel, Shire chief executive, quoted in *Financial Times*, 22 May 2004.

Table 4.2. Pharma/biotech acquisitions, 2003–2008

Acquirer	Target	Upfront deal value ($m)	Overall deal value ($m)	Date
Chiron (US)	PowderJect	810	810	2003
UCB (Belgium)	Celltech	2,700	2,700	2004
AstraZeneca (UK)	KuDOS	210	210	2005
Sosei (Japan)	Arakis	186	186	2005
AstraZeneca (UK)	Cambridge Antibody	1070	1070	2006
Emergent Biosolutions (US)	Microscience	NA	NA	2005
Novartis (Switzerland)	Neutec	586	586	2006
Pfizer (US)	PowderMed	300	400	2006
GlaxoSmithKline (UK)	Domantis	454	454	2006
AstraZeneca (UK)	Arrow	150	150	2007
Takeda (Japan)	Paradigm	NA	NA	2007
Sanofi (France)	Acambis	550	550	2008
Roche (Switzerland)	Piramed	160	175	2008
Wyeth (US)	Thiakis	30	150	2008

Note: After 2000 there was a trend in pharma/biotech acquisitions for part of the purchase price to be paid at the start, with the rest depending on the progress of the drugs being developed by the acquired firm.
Source: HBM Pharma-Biotech M & A Report.

UCB. But the deal formed part of a trend which was to become a source of anxiety among policy makers: the acquisition of promising biotech firms by Big Pharma companies, most of which were non-British (Table 4.2).

The most aggressive acquirers were American; one commentator referred to a 'feeding frenzy', as US companies picked off attractive European targets in biotech and related sectors.[30] The largest of these deals was General Electric's $5.7bn takeover of Amersham. This company (once part of the Atomic Energy Authority, privatized in 1982) was a diagnostics rather than a drug discovery firm, and since GE planned to make it the centre of a new healthcare division the UK was likely to gain more than it lost from the transaction. For investors, nevertheless, Amersham had been an important constituent in their healthcare portfolios, and its removal might reduce their interest in the sector.

As for the government's reaction to the Celltech sale, the UK had a long-established policy of openness towards inward investment. Although Fellner received some anxious calls from the Department of Trade and Industry when the deal was announced, there was no attempt on the UK government's part to discourage the merger. There was a broad consensus across the political spectrum that, on balance, foreign acquisitions were good for the economy because the new owners brought in capital, technology, and management and were likely to add value to what they bought. In the case of Celltech this optimistic view seemed plausible, since UCB, although not the blue-chip

---

[30] *Financial Times*, 12 November 2003.

pharmaceutical company that Fellner and his colleagues might have hoped for, had every intention of using Celltech's Slough R&D centre as the core of its planned push into biotech (and UCB continues to operate the site in 2015). Celltech's CDP870, given the trade name Cimzia, suffered several delays in development, including requests for additional data from the FDA, and it was overtaken by other drugs in the race for what became the multi-billion dollar anti-TNF market.[31] But Cimzia was eventually launched in the US in 2008 and in Europe in 2009; in 2014 the drug achieved sales of €797m.

Shortly before the UCB deal another promising firm, PowderJect, had been snapped up by an American company. PowderJect was a spin-out from Oxford University, based on a novel approach to vaccination which involved sending small particles through the skin without the use of needles. Floated in 1997, it was run by Paul Drayson, an ambitious non-scientist entrepreneur whose earlier experience had included spells with the automotive manufacturer, British Leyland, and with Trebor, the confectionery company; he had organized a management buy-out of one of Trebor's businesses with the help of 3i. Searching for another new business to start, he joined forces with Professor Brian Bellhouse, the Oxford scientist who had invented the technology on which PowderJect was based.[32]

After 2 years in the job Drayson reached the conclusion that the right strategy for PowderJect was not simply to sell its technology in return for a royalty but to develop and make vaccines itself. Some critics complained that he was turning his back on an innovative technology in favour of a 'plain vanilla' business, but the change of direction worked well. He bought from Celltech the ex-Medeva vaccine plant at Speke, acquired the leading Swedish vaccine producer, and came close to a merger with a large German company. By the end of 2002 PowderJect achieved the distinction, rare in British biotech, of making a profit.

PowderJect was now the sixth largest vaccine company in the world, and a likely acquisition target for other producers which were looking for scale and increased market share. When Chiron, the US biotech firm, announced its intention to make a bid for the company in November, 2002, Drayson persuaded his shareholders to reject it, but the offer was subsequently raised to a level which was impossible to resist. The bid valued PowderJect at just over £500m.[33] Several of Drayson's colleagues moved on to senior positions in

[31] Cormac Sheridan, 'Cimzia's setback paves way for other TNF inhibitors in Crohn's disease', *Nature Biotechnology* 25 (2007) pp. 487–8.

[32] Douglas Hague and Christine Holmes, *Oxford Entrepreneurs* (Oxford: Said Business School, 2006) pp. 87–90.

[33] In 2004 Chiron spun off PowderJect's needle-free delivery technology into a separate company, PowderMed. Two years later PowderMed was acquired by Pfizer. Chiron itself was acquired by Novartis in 2005, marking yet another change of ownership for the Speke vaccine plant. In 2014 Novartis sold its flu vaccine business, including Speke, to CSL of Australia.

other biotech firms, while Drayson himself later entered politics, joining the Labour government in 2005 as Minister for Defence Procurement and subsequently serving as Minister for Science (see Chapter 8).

The Celltech and PowderJect deals were welcomed by investors, relieved to see that some managers of UK biotechs were conscious of the need to generate shareholder value. There was also a beneficial side-effect. In the US the recycling of talent that followed mergers and acquisitions had created a pool of experienced managers and scientists that younger firms could draw on; the same thing happened, on a smaller scale, in the UK. Steve Harris, PowderJect's chief financial officer, helped to found Zeneus, a speciality pharmaceutical company which was formed in 2003 and sold 2 years later to Cephalon of the US for $360m.[34] Harris was later a co-founder of Circassia, a developer of anti-allergy therapies (discussed in the next chapter).

In the Celltech case, several managers left after the UCB deal to join an early-stage firm, Chroma.[35] Raj Parekh, who had worked for OGS as chief scientific officer and then for Celltech after the merger, became chairman of Chroma.[36] One who stayed with UCB was Melanie Lee, who had joined Celltech from GlaxoWellcome in 1998 as head of research and development; she worked for UCB until 2009, working closely with Roch Doliveux, chief executive, in building up the Belgian company's biotechnology business.[37]

Nevertheless, the Celltech and PowderJect acquisitions underlined the fact that more than 20 years after Celltech's founding, the UK had yet to produce an independent biotech firm that could stand comparison with what Genentech and others had achieved in the US. This was a matter of concern in government, and not just because of nationalism, or envy of the US. The worry was that the British life sciences industry had become over-dependent on the two Big Pharma companies, GlaxoSmithKline and AstraZeneca.[38]

---

[34] The principal founder of Zeneus was Bryan Morton, an ex-pharma executive who became a serial entrepreneur. After Zeneus he went on to found another speciality pharma company, EUSA Pharma, which was sold to Jazz Pharmaceuticals of the US in 2012 for $650m. Morton later became chairman or director of several biotech firms.

[35] Chroma Therapeutics, founded in 2000 with backing from Abingworth, used chromatin biology to develop treatments in the fields of oncology and inflammatory disorders; the two scientific founders were Paul Workman from the Institute of Cancer Research and Tony Kouzarides from Cambridge. Chroma raised Series B and C funding in 2004 and 2006 but plans for an IPO were frustrated by the world financial crisis. It later turned itself into a 'virtual' firm, drawing income from the partnerships it had formed with pharmaceutical companies.

[36] In 2005 Parekh joined Advent, the venture capital firm, as a general partner.

[37] After leaving UCB Melanie Lee was chief executive of two start-up firms before joining BTG as chief scientific officer in 2015.

[38] During the 1990s the world pharmaceutical industry had been going through a process of consolidation, involving several large mergers and acquisitions on both sides of the Atlantic. Two of the smaller British firms, Boots and Fisons, had sold their pharmaceutical businesses to foreign acquirers. Glaxo acquired Wellcome in 1995, and the enlarged group merged with SmithKlineBeecham in 2000 to form GlaxoSmithKline. Zeneca, demerged from ICI in 1993, merged with Astra of Sweden in 1999.

It needed a tier of profitable, medium-sized biotech firms beneath the two giants.

A Potential Star: Cambridge Antibody Technology

If innovative science was the key, or at least one of the keys, to becoming a world leader, then Cambridge Antibody Technology seemed capable of achieving that status. CAT was a pioneer in the development of fully human monoclonal antibodies, based on its proprietary phage display platform; this technology made possible the isolation of antibodies that would bind strongly to target antigens without provoking the adverse immunogenic response associated with murine-based monoclonal antibodies.[39]

CAT had a difficult start, struggling to make its technology work and using licensing and partnerships to help finance its research. One of several deals signed in the early 1990s was with Knoll, a subsidiary of BASF in Germany. This was a collaboration under which CAT was required to deliver an optimized antibody candidate against targets selected by Knoll; the chosen molecule was an anti-TNF antibody for the treatment of rheumatoid arthritis—the same target that Celltech was aiming at with CDP870.

The German company paid CAT's research costs and CAT was entitled to milestone payments in the course of development and a royalty if the chosen molecule reached the market. That molecule, adalimumab, was to lead, 10 years later, to Humira, a humanized monoclonal anti-TNF antibody for the treatment of autoimmune diseases, which became the world's best-selling drug—in 2013 revenues from Humira exceeded $10bn. No one anticipated that outcome when the licensing deal was signed.

By 1997, when CAT had its initial public offering on the London Stock Exchange, it had decided to reduce its dependence on partnerships and to develop its own antibody-based drugs. Whether this was the correct strategy had been a contentious issue on the board before the flotation, with some directors, including Greg Winter, questioning the wisdom of shifting away from acting as partner and technology provider to Big Pharma; there were also questions as to whether the drug candidates chosen for in-house development were the right ones. Winter left the board shortly before the flotation.

Progress in implementing the strategy was slower than expected, and in 2002 the Board decided, with the agreement of the chief executive, David

---

[39] An alternative approach to the development of human monoclonal antibodies was based on transgenic mouse technology, used by Abgenix and Medarex in the US. CAT's European rival in phage display was Morphosys in Germany; the two firms had a lengthy patent dispute which was finally settled in 2002.

Chiswell, that a change at the top was needed. Chiswell was succeeded by Peter Chambré, who had been chief operating officer at Celera Genomics, a US biotech firm; most of his earlier career had been in marketing and general management in the UK.[40]

At the time of Chambré's appointment the company was 'burning' some £40m–£50m a year in research expenditure; it had enough cash to keep going for another 2–3 years. With UK biotech shares in the doldrums, there was no possibility of raising more money from investors. One positive factor was the imminent approval by the FDA of Humira, which was now in the hands of Abbott in the US (BASF having sold its pharmaceutical business to Abbott in 2000). CAT was entitled to a royalty on Humira sales, but income from this source was not expected to be enough on its own to generate the cash that CAT needed for its own drug development programmes.

To speed up CAT's transition from technology provider to drug developer, Chambré's plan was, first, to make acquisitions, preferably of firms that had late-stage drugs near to commercialization, and, second, to negotiate one or more partnerships on better terms than the licensing deals on which CAT had previously relied. In January 2003, as noted earlier, CAT announced an agreed merger with Oxford GlycoSciences; but the deal fell apart after a drop in CAT's share price, caused by an unexpected request from Abbott that the royalties payable to CAT on Humira should be reduced to just 2 per cent in order to offset the payments that had to be paid to other firms that had participated in the development of the drug.

When CAT's general counsel reported this unwelcome news to the Board, it was clear that if Abbott intended to dispute the royalty agreement this would have to be disclosed to shareholders. The company sought clarification from Abbott, and the reply was equivocal. When CAT informed shareholders that the company was potentially in dispute with Abbott over Humira, the share price fell to just over £3, compared to over £5 when CAT's original offer had been made. This left the OGS board no option but to recommend the rival offer from Celltech.

CAT now had to deal with the Abbott situation and, since the takeover route was likely to be blocked until the dispute was resolved, to find some other way of driving the company forward. Chambré and his team believed that, given its exceptionally strong science, CAT should be capable of securing a different sort of partnership from those that it had negotiated in the past. The model in his mind was the agreement that Genentech had made in 1990 with Roche, whereby the Swiss company had provided cash and had

---

[40] Chiswell later served as chairman or director of several biotech firms, including Kymab, a Cambridge-based monoclonal antibody firm.

taken a large shareholding, but had left Genentech as a separately listed and largely autonomous company.

CAT approached six companies in the course of 2003; by the spring of 2004 the runners were reduced to two, and eventually to one, AstraZeneca. Astra-Zeneca had been slower than other Big Pharma companies to explore the possibilities offered by novel antibody therapies, and was still largely dependent on the chemistry-based approach to drug discovery. Its management saw a partnership with CAT as a way of establishing a stronger position in biologics.

Under the 5-year deal signed in November 2004 the two companies agreed to work together on an antibody-based pipeline in several therapeutic areas, excluding cancer, where CAT would maintain its internal development pipeline. AstraZeneca put £75m into the company and acquired 20 per cent of its shares. The AZ deal, as Chambré said later, 'transformed the future economics of CAT, although not its immediate profit prospects—we now had a Genentech/Roche type arrangement'.[41]

Attempts to settle with Abbott had not been successful and litigation commenced in the High Court in London in November. In December the court found in CAT's favour and Abbott appealed. In October 2005, shortly before the appeal was due to be heard, CAT announced a settlement which provided some improvement for CAT's royalty payments.

The challenge now was to build on the AstraZeneca deal and to convert CAT into a profitable product-based company. The possibility that AstraZeneca might want to buy the remaining 80 per cent was not explicitly on the agenda when the agreement was signed, but as the partnership developed senior managers in the pharmaceutical company began to favour a closer relationship. Several methods were considered, including an arrangement whereby CAT would continue to have a separate stock market listing, but AstraZeneca ultimately decided to go for a full takeover and this was agreed by CAT shareholders in July 2006.

AstraZeneca paid £13.20 a share for the shares it did not already own, valuing the whole company at £702m; the price was pitched at a 67 per cent premium to the pre-bid price.[42] The value of the deal for the acquirer lay not in Humira (shortly after the takeover, AstraZeneca sold the Humira royalty stream to Royalty Pharma of the US for £416m), but in CAT's monoclonal antibody platform, which was expected to rejuvenate AstraZeneca's drug pipeline.[43]

---

[41] Interview with Peter Chambré, 5 December 2013.

[42] Earlier in 2006 Amgen had paid $2.2bn for Abgenix, a US rival antibody technology developer, offering a premium of 51 per cent on the pre-bid share price.

[43] Royalty Pharma was not a drug developer, but specialized in acquiring revenue-producing intellectual property from other companies.

In 2007 AstraZeneca greatly increased its commitment to biologics when it bought MedImmune, a large US biotech, for $15.6bn. MedImmune had been formed in 1988, the year before CAT, to exploit humanized monoclonal antibodies. It had grown faster than CAT, mainly because it had been quicker to bring drugs to the market; its biggest-selling drug, Synagis (palivizumab), an antibody-based treatment for respiratory infections, was launched in 1998. Whereas CAT's strength lay at the upstream end of the drug discovery process—the manipulation and engineering of antibodies to maximize their therapeutic properties—MedImmune was stronger in clinical development and commercialization; in that sense the two acquisitions were complementary.

AstraZeneca was criticized for paying too much for MedImmune, and several of the US firm's drug candidates were not successful. But the two acquisitions gave AstraZeneca access to research and manufacturing capabilities that it did not possess. The proportion of biologics in AstraZeneca's pipeline grew from less than 10 per cent at the time the deals were done to about 50 per cent in 2015. The CAT laboratories were expanded after the acquisition and in 2013, as part of a wide-ranging reorganization, AstraZeneca announced plans to move its main UK research centre from Alderley Park in Cheshire to Cambridge.[44]

Thus the acquisition of CAT by AstraZeneca ensured that the pioneering research of Greg Winter and his colleagues was put to good use. On the other hand, like the UCB/Celltech deal 2 years earlier, the disappearance of CAT as an independent company removed a potential beacon from the UK biotech sector.

Could that outcome have been avoided if UK investors had been more supportive? This question acquired greater salience over the next few years in the light of the spectacular success of Humira; more of the profits might have come back to the UK if CAT had stayed independent. However, when BASF negotiated the licensing arrangement in 1993, CAT—which had struggled to attract investment in the sparse UK funding environment of its early years—was not in a position to demand a greater share of whatever profits might flow from it.

It is conceivable that CAT might have stayed independent for longer if it had had one or more large, patient investors who believed that in the long run a stand-alone company would achieve greater returns than were available from a trade sale. But that would have required a degree of confidence in CAT's pipeline that was probably unrealistic. As it was, the CAT board recommended acceptance of the price offered by AstraZeneca, and there was no opposition to the sale from shareholders.

---

[44] AstraZeneca's other main research centres were at the ex-MedImmune site at Gathersburg, Maryland, and the ex-Astra site at Mölndal, Sweden.

De-risking Strategies

For Celltech or CAT to have stayed independent would have required capital on a scale that UK investors would have been unlikely to provide, given the uncertain future of their research programmes. Were there other, less risky ways of surviving in the drug development business?

An alternative model was that followed by Shire, which had begun by establishing sales from licensing-in approved drugs before building a portfolio of late-stage drug candidates. In 2000, following the takeover of Roberts, Rolf Stahel negotiated the $4bn purchase of a Canadian company, Bio-Chem Pharma, which had an influenza vaccine on the market and a stream of royalty revenues; it also had a biologics facility that was developing recombinant protein vaccines against bacterial infections. The next step, which came after Stahl had been succeeded as chief executive by Matthew Emmens, was the purchase of a Boston-based company, Transkaryotic Therapies (TKT), which had a portfolio of recombinant protein-based treatments for rare diseases caused by enzyme protein deficiencies. Some shareholders were uneasy about this deal because it took Shire nearer to high-risk early-stage research but Emmens convinced them that the deal would diversify the company's revenue base as well as providing opportunities for cost savings. Further acquisitions over the next few years included Jerini in Germany and Movetis in Belgium, making Shire an even more global company, and even less connected to the UK.[45]

These deals reduced Shire's dependence on the ADHD drugs on which it had relied in the early years and strengthened its position in treatments for orphan diseases. Shire's criteria for building a successful speciality pharma company were set out by Emmens in 2007: 'We want to access products that have sales potential in the $200m–$400m a year range and are unlikely to be on Big Pharma's radar, would require a small sales force and have low development risk.'[46]

Shire's approach was admired by Louise Makin, a former Baxter executive who was appointed chief executive of BTG in 2004.[47] Under Makin's leadership, BTG evolved from its original métier as a technology licensing business into a drug development company. Like Shire, it looked for niche markets and it steered clear of early-stage research. In 2008 BTG spent £218m to acquire Protherics, a British firm which had several polyclonal antibody drugs on or near the market, including CytoFab, a treatment for sepsis, as well as a US

---

[45] Although Shire retained its development facility in Basingstoke, its links with the UK were further reduced in 2008 when for tax reasons it moved its headquarters to Ireland.

[46] *BioCentury*, 25 June 2007.

[47] Louise Makin studied materials science at Cambridge, and then worked for 13 years in ICI. She served as director of global ceramics in English China Clays before joining Baxter in 2000.

marketing organization.[48] Two years later BTG paid £177m to acquire Bio-compatibles, which made medical devices such as stents and microbeads used to deliver oncology drugs; the purchase was in line with Makin's focus on high-margin products sold to specialist physicians.[49]

There were other ways of reducing risk. When Peter Fellner at Celltech bought Medeva in 1999 he was seeking to offset the costs of early-stage research by acquiring revenue-generating businesses. In 2003, having taken on the post of non-executive chairman of British Biotech in the previous year, he repeated the formula. Together with an ex-Celltech colleague, Simon Sturge, he engineered a three-way merger between British Biotech, Ribotargets, and Vernalis. British Biotech, the one-time flagship of the industry, had cash but not much of a pipeline.[50] Ribotargets was a Cambridge-based drug discovery firm spun out of the Medical Research Council. Vernalis (formerly Vanguard Medica) had a migraine drug, frovatriptan (Frova), launched on the market in 2002.[51] Sturge, chief executive of the enlarged group (which took on the Vernalis name), said that further acquisitions were likely, probably of firms that had late stage products and revenue streams. 'We are not looking for technology or early-stage research companies', he said.[52]

One of the biggest venture rounds in 2004 was for Arakis, a company backed by Merlin and founded by former employees of Chiroscience. Andy Richards, who had left Chiroscience shortly before the Celltech takeover, was approached by two former colleagues, Julian Gilbert and Robin Bannister, who planned to commercialize a re-profiling technology—identifying new uses for off-patent drugs—on which they had worked at Chiroscience. That this was a lucrative business model was demonstrated in 2005 when Arakis, in partnership with Vectura, a drug delivery firm, signed a development deal with Novartis worth up to $375m in milestone payments and royalties, on a new treatment for chronic obstructive pulmonary disease.[53] Arakis was bought a few months later for £107m by Sosei of Japan, a satisfactory return for its owners only 3 years after the company had been founded.[54]

---

[48] The polyclonal antibodies that Protherics had developed were based on the long established process for making antiserum by raising and extracting antibodies from animals exposed to antigens, rather than through monoclonal antibody technology; Protherics had been formed in 1999 by a merger between Proteus International and Therapeutic Antibodies.

[49] *Financial Times*, 20 November 2010.

[50] Eliot Goldstein, who had taken over as chief executive of British Biotech after the resignation of Keith McCullagh in 1998, had pursued various ways of reviving the business, including a possible merger with Morphosys of Germany. When those talks broke down at the end of 2002, Goldstein resigned and Peter Fellner took on the role of non-executive chairman.

[51] The drug, marketed by Endo (a US speciality pharma company) as Frova, is a traditional small molecule drug that had been licensed in from SmithKline and French. Developed for menstrually associated migraine, it became a niche product with relatively low sales.

[52] *BioCentury*, 14 July 2003.   [53] *Financial Times*, 13 April 2005.

[54] Like several other Japanese pharmaceutical companies, Sosei was using acquisitions in Europe and the US as a means of reducing its dependence on its domestic market (see Chapter 6).

Vectura, Arakis's partner in the Novartis deal, specialized in inhaled drugs, and, like Arakis, did not engage in drug discovery.[55] Chris Blackwell, chief executive, explained his strategy. 'We would not develop drugs where there were any novel, active components but we would pursue a strategy of repurposing—taking a drug that had already secured regulatory approval and was on or near the market, and repurpose it, either by developing a different route to delivery or by reformulating it so that it could be used in a different disease indication.'[56] This was the kind of business that generalist investors liked. Vectura raised £20m when it listed on AIM in 2004, and moved to the main market in 2007.

In the financial conditions that prevailed in the early 2000s de-risking strategies of this sort made sense. But the biggest rewards in biotech, as US experience had shown, came from breakthrough medicines based on innovative science. This was what the few UK-based venture capitalists that were still interested in start-ups were looking for, and they were prepared to back such firms as long as they had confidence in the science, the management, and the business plan. There were also some US venture capitalists which were seeking that sort of opportunity in the UK. In 2005 MPM Capital from the US led a £32m fund raising round for Oxagen, which had shifted from its earlier focus on genetics to build a pipeline around a family of drug targets known as GPCRs (G-protein coupled receptors); its lead drug was a treatment for asthma.

MPM and the other VCs which took part in the round probably assumed that the most likely exit for Oxagen was a trade sale.[57] Yet there were signs at that time that the alternative exit route, the IPO, might once again be feasible. In the US, after a 3-year lull following the bursting of the genomics bubble, investor interest in biotech had revived and the IPO window on NASDAQ had reopened. Several British firms, hoping that the change in sentiment might spread across the Atlantic, began to prepare for an IPO in London.

The London IPO Market

The first to break the ice was Ark Therapeutics, a gene therapy firm which had come out of University College London with backing from Merlin. When it floated on the main market in March 2004 Ark claimed that it was not an

---

[55] Vectura was founded in 1997 with seed funding from Merlin Ventures.

[56] Interview with Chris Blackwell, 18 September 2013. Chris Blackwell stepped down as chief executive in 2015 and was replaced by James Ward-Lilley, a senior AstraZeneca executive.

[57] Oxagen's asthma programme ran into difficulties and in 2012 the technology was licensed to a Russian company, Eleventa. At the same time Oxagen created a new firm, Atopix, to develop a treatment for atopic dermatitis.

early-stage biotech since it had three lead products in advanced clinical trials and a fourth, a wound dressing device, which was already on the market. Its most promising asset was Cerepro, a gene-based therapy for brain tumours which had been given orphan drug status by the FDA and the European Medicines Agency. The £55m offering was three times over-subscribed, but by September, even though the company claimed to have delivered on all its promises, the shares had fallen from a peak of 141p just after the IPO to 70p.[58]

No other British company floated on the main market in 2004. Only two, ProStrakan and Ardana, did so in 2005, and Renovo followed in 2006. Pro-Strakan's business model was similar to Shire's, based on in-licensed drugs rather than drug discovery.[59] Ardana had been founded in 2001 to commercialize research from the MRC's reproductive sciences unit in Edinburgh. Renovo, spun out from Manchester University by Mark Ferguson and Sharon O'Kane, was developing drugs to prevent skin scarring after surgery; it had raised £31m in two rounds of funding before the IPO.

The 2004–07 period also saw several biotech flotations on AIM. London's junior market had come through the 2000–01 boom-and-bust in better shape than its European competitors; Germany's Neuer Markt, founded in 1997, had collapsed in 2002 after a series of disasters among newly listed firms (see Chapter 6). The authorities in the London Stock Exchange claimed that AIM was now the logical base for a pan-European high-technology market. Yet the amounts of money that could be raised on AIM were small by biotech standards, generally in the £10m–£40m range. For venture capital firms an AIM listing rarely provided a complete exit; they generally continued to hold shares in the listed firm and were expected to contribute to follow-on capital issues.[60] Only in a few cases did AIM companies graduate to the main market.

Drug discovery firms which had large ambitions saw AIM as a poor relation compared to the main market. But there was a deeper problem. The US, thanks to the size and growth of the biotech sector, had fostered an investment community that was big enough to support an array of specialist healthcare investors and expert biotech analysts. That investment infrastructure was missing in the UK. Would British firms get a better reception if they floated in New York rather than London?

---

[58] *Financial Times*, 1 September 2004.

[59] ProStrakan had been created in 2004 by a merger between Strakan, a Scottish company whose founder, Harry Stratford, had been a co-founder of Shire, and Proskelia, a French company that had been spun out of Aventis.

[60] Pre-IPO investors in newly floated public companies are required to observe a lock-in period during which they cannot sell their shares for a set period of time. Given the small pool of investors in junior markets, conditions may not allow existing shareholders to sell large stakes without bringing down the share price substantially, so lowering their returns.

The Lure of the US

In the early 1990s, before the London Stock Exchange changed its rules to accommodate loss-making biotech firms, several UK companies had listed their shares on NASDAQ, but the results were generally disappointing. Their experience had shown that if a non-American biotech firm was to succeed in the US it had to spend time and effort on cultivating US investors, prepare the ground well, and have a good story to tell.

One possible route was to buy an American company with an established NASDAQ listing. In 2005 Cyclacel, a Dundee-based firm which had been considering an IPO in London, engineered a reverse takeover of a NASDAQ company, Xcyte Therapies, and shifted its headquarters to New Jersey; the main research facility remained in Dundee. XCyte had cash that would be useful as Cyclacel's lead drugs entered clinical trials, and a further $45m was raised from US investors in 2006. The NASDAQ listing, according to Cyclacel's chief executive, Spiro Rombotis, gave the company 'a sophisticated US shareholder base, better access to a large number of savvy biotech fund managers, and closer proximity to an experienced pool of management talent'.[61]

Biovex, which had been spun out from University College London in 1999, had raised some $50m in a series of funding rounds and by 2004 was also considering an IPO in London. In March of that year the finance director, Philip Astley-Sparke, put to his colleagues a proposal for relocating the company to the US and listing on NASDAQ.[62] He pointed out that the company's lead programme, OncoVEX, a novel treatment for melanoma, was about to enter Phase II trials in the US; clinical investigators in the US had more experience with this type of drug than in Europe and patient recruitment was likely to be easier. In addition, the financing risk for Biovex would be reduced if it moved to the US, since US investors were more willing than their European counterparts to fund novel technologies.

The first reaction of the Biovex board was sceptical, but they accepted the force of the argument and the move went ahead in 2005. A biologics manufacturing plant was built near Boston, benefiting from the specialist expertise which was available locally in the design of such facilities; the research laboratories remained in Milton Park, near Oxford.

The next few years were not easy for Biovex—an attempt to float on NASDAQ failed and several hoped-for partnerships did not materialize—but in 2011 the company was sold to Amgen for up to $1bn ($425m up front and a further potential $575m linked to milestones). Amgen retained the research laboratory in the UK. In April 2015, following a successful Phase III trial, the

---

[61] *BioCentury*, 19 December 2005.
[62] Interview with Philip Astley-Sparkes, 9 June 2014.

lead drug, now referred to as T-VEC (short for talimogene laherparepvec) and given the trade name Imlygic, was approved by an FDA advisory committee; final approval was expected later in the year. The outcome was a triumph for UK-originated science, though one that brought little profit for the original investors in the company.[63]

A later emigrant to the US was Astex, which had a proprietary platform for fragment-based drug discovery, based on x-ray crystallography and other computational techniques; it was widely regarded as a world leader in that field. This company was founded in 1999 by Harren Jhoti, a senior scientist in GlaxoWellcome, and two Cambridge professors, Tom Blundell and Chris Abell.[64] By 2006 Astex had made enough progress to contemplate an IPO in London, but internal problems arising from the resignation of the chief executive caused a postponement, and the IPO opportunity disappeared with the onset of the world financial crisis. The company was loss-making and short of cash, forcing Jhoti—now chief executive, having previously been chief scientific officer—to reduce costs, concentrate on a few priority drug programmes, and look for partners, some of whom might also be potential acquirers.

At this point Jhoti received an unexpected merger proposal from a NASDAQ-listed American company, SuperGen. SuperGen's chief executive, Jim Manuso, pointed out that his company was already earning revenue from a marketed drug and had another compound in late stage development; what it lacked was a discovery platform to feed a longer term development pipeline, and that was what Astex could provide. Despite initial misgivings among the Astex directors, Jhoti saw that the proposal made sense. 'I told the Board not to think of the SuperGen deal as an exit but as a financing event. It would allow us to become a public company in the US, putting ourselves into the US window; we would get access to about $300m in licensing revenue; and it would give investors liquidity.'[65]

The merger went through in 2011 (the enlarged company kept the Astex name) and for a while the two sides worked together harmoniously; the business was divided between the R&D facility in Cambridge, and the commercial side in the US. However, by 2013 a combination of factors, including dissension on the board about leadership and about strategic direction, led to

---

[63] Few of the early investors in Biovex made much money out of the Amgen deal; their shareholdings had been diluted by investors who came in later. One of the academics who co-founded Biovex, David Latchman from UCL, reported in 2014 that his share of the amount paid by Amgen was £423. Letter to *Times Higher Education Supplement*, 3 July 2014.

[64] Sir Thomas Blundell was head of the Department of Biochemistry in Cambridge, having previously worked with Jhoti in the Department of Crystallography at Birkbeck. Between 1994 and 1996 he was chief executive of the Biotechnology and Biological Sciences Research Council. Chris Abell was a professor in the Department of Chemistry.

[65] Interview with Harren Jhoti, 11 February 2014.

the decision to put the company up for sale; it was bought by Otsuka of Japan for $886m.

For Otsuka's president, Taro Iwamoto, the main attraction of Astex lay in the Cambridge research capability. It was a way of migrating his research base to the West, but the sale almost certainly would not have happened if Astex had not moved to the US. 'It put us in the US shop window', Jhoti says; 'we got covered by analysts and acquired a stronger investor base—and we gained the resources to do the clinical development more effectively'. There was every prospect that Astex's UK-based research activity would be maintained and expanded under Otsuka's ownership.

Asked whether Astex could have remained a British company, Jhoti replied: 'I was trying to build Astex to be the next Vertex, a company I admired—I liked the way that company was set up and pioneered a new approach to drug discovery, I wanted to emulate them. I would have liked to IPO in London but it has been very difficult to get traction in the City for biotech stocks.'[66]

Benefits of a US Connection

In these cases the change of domicile did not directly affect the research function, which remained in the UK. In that sense the outcome could be seen as positive, illustrating the benefits of linking British science to the commercial and financial strengths of the US. For British policy makers, however, the shift to the US posed two dangers: that discoveries made in the UK would be exploited mainly in the US, just as penicillin had been; and that any hope of creating a viable biotech sector in the UK would be frustrated if the best firms moved across the Atlantic.

These issues were highlighted by the case of Solexa, a Cambridge-based gene sequencing firm, which came to be seen by some observers as a classic example of how a brilliant piece of innovation by British scientists was 'lost' to the US.

Backed by Abingworth, Solexa was set up in 1999 by two Cambridge scientists, Shankar Balasubramanian and David Klenerman, to develop an approach to DNA sequencing which promised to be faster and cheaper than existing methods.[67] They were building on work which had started in the 1970s with the invention by Frederick Sanger, a Cambridge biochemist, of the chain-terminal method of sequencing DNA molecules.[68] After 4 years of

[66] Ibid.
[67] Other investors included Amadeus Capital Partners, one of the leading Cambridge venture capital firms, which led the $14.4m Series B financing in 2004.
[68] Frederick Sanger was one of the few scientists to win two Nobel prizes. The Sanger Institute, set up in Cambridge in 1993 with support from the Wellcome Trust and the Medical Research Council, became a leading centre of genome research.

experimentation Solexa ran into a technical problem. The single-molecule approach to sequencing which Balasubramanian and Klenerman had adopted did not produce a strong enough signal. The solution was found in a technology developed by a Swiss firm, Manteia, which created clusters of identical molecules of DNA. Manteia was close to bankruptcy and its owner, Serono, had put it up for sale.[69]

Solexa made an offer for the business, but found itself in competition with a NASDAQ-listed sequencing firm, Lynx Therapeutics. The two companies agreed to make a joint offer, which was successful, and their cooperation in this deal led to the idea of a merger, with Lynx using its shares to acquire the British company; the merger went through in 2005. The NASDAQ listing allowed Solexa to raise new capital and to get greater visibility in the US, which was by far its biggest market. A new chief executive, John West, an expert in instrumentation for gene sequencing, was recruited from Applied Biosystems, one of Solexa's main competitors.

Access to capital and management was the principal reason for the move to the US; another factor was the depth of instrumentation engineering skills that was available in the San Francisco Bay Area. What Solexa also needed was a distribution network to serve American and international customers. For Solexa to have built such a network on its own would have been prohibitively expensive, hence the decision in 2006 to accept a $650m takeover offer from Illumina, a US company. Illumina, founded in 1998 in San Diego (just a year earlier than Solexa), had already developed a range of life science tools and technologies, including an innovative system for rapid, large scale genotyping, that served the same customers as Solexa. The acquisition of Solexa paved the way for a new generation of DNA sequencing instruments which were crucial to Illumina's subsequent growth. It continued to rely on Solexa's Cambridge-based researchers.[70]

'Going to NASDAQ via Lynx and raising $100m was crucial', says Nick McCooke, who was chief executive at the time the Lynx merger was negotiated; 'what we were doing was totally novel—there was no way we could have raised the money we needed in Europe'.[71] David Klenerman, one of Solexa's scientific founders, said: 'The Illumina deal made considerable business sense. It got the product to the market much quicker and this was a key aspect of its success.'[72] Stephen Bunting of Abingworth, who regards Solexa as the single most successful example of technology transfer in the UK, said: 'The UK had

[69] Kevin Davies, *The $1,000 Genome* (New York: Free Press, 2010) p. 107.

[70] In 2012 Illumina was the subject of a $6.7bn takeover bid from Roche, which was mainly interested in the sequencing technology, but the bid was successfully resisted despite the premium of over 60 per cent to the pre-bid share price.

[71] Interview with Nick McCooke, 5 February 2015. McCooke was one of the many UK biotech executives who had spent part of his career with Celltech.

[72] Email communication from David Klenerman, 25 October 2014.

the science, all the innovative knowhow was done in Cambridge—that was where those skills were, but when it came to commercialisation, finance and engineering the US had the edge.'[73]

Trade Sales Continue

With or without a NASDAQ listing, a presence in the US was increasingly seen as essential for any ambitious biotech firm. When Greg Winter and Ian Tomlinson in the Laboratory of Molecular Biology founded Domantis in Cambridge in 2000 (this was Winter's second start-up, following Cambridge Antibody Technology), they appointed an experienced American manager, Bob Connelly, as chief executive, and the company's commercial and financial headquarters was set up in Waltham, Massachusetts, close to the Boston biotech cluster.[74] Research remained in Cambridge.

By 2005 Domantis had a portfolio of pre-clinical assets based on its unique technology for producing a new generation of much smaller antibody-derived drugs (one tenth of the size of normal antibodies) that it was hoped would be more therapeutically effective. The approach was seen in the industry as having great promise, and Domantis negotiated six partnerships with pharmaceutical companies and biotech firms, mainly in the US. To take these assets into clinical trials, and to build up manufacturing capacity, called for new funding. One possibility was to do a deal with Big Pharma similar to the one that Roche had done with Genentech in 1990 and that CAT had tried to imitate with AstraZeneca. But this would have been a more uneven partnership; Domantis was a small, loss-making firm, whereas Genentech at the time of the Roche deal was already a well-established company, and CAT was poised to benefit from a substantial royalty stream. Another option was to convert Domantis into an American company as Cyclacel had done, list it on NASDAQ, and then raise new money.

While these possibilities were being considered, Domantis had an informal takeover approach from an American pharmaceutical company, which saw the acquisition of Domantis as a way of moving into the next generation of monoclonal antibodies; the indicated purchase price was around $500m. As Ian Tomlinson recalled later, 'it was a jaw-dropping moment for Bob Connelly and me—we had to take it seriously'.[75] Through its US advisers, Lehman Brothers, Domantis approached other companies and by October 2006 it

---

[73] Interview with Stephen Bunting, 3 December 2013.
[74] As in the case of CAT, the lead investor in Domantis, with just over 30 per cent of the shares, was Peptech, the Australian biotech firm.
[75] Interview with Ian Tomlinson, 15 October 2013.

had received offers, all pitched around $500m, from four pharmaceutical companies and one large biotech firm.

The front runner was GlaxoSmithKline, not just because it was British and had its main R&D base at Stevenage, not far from Cambridge, but also because the newly appointed GSK research director, Moncef Slaoui, had a clear view of why GSK wanted Domantis and of how it would be integrated into the larger organization. Agreement was reached at the end of 2006, with GSK paying £230m for a company which was less than 7 years old. The sale returned to investors more than five times the money invested, making Domantis a faster and more lucrative exit than had been achieved by Cambridge Antibody Technology when it was bought by AstraZeneca.[76] The purchase of Domantis gave GSK access to a novel monoclonal antibody technology as well as the skills of some sixty trained antibody engineers, molecular and cell biologists; it also provided the impetus for the creation of a new Biopharmaceutical R&D unit in Cambridge.

Did Domantis have to be taken over? As Tomlinson recalled, 'before the formal GSK offer came in there was a lot of internal discussion over whether we should take it. It was fairly finely balanced—we all knew that if the offer was real there would have to be a very compelling case not to take it. There was also the argument that in the end, if you want to make medicines, you have to hook up with a pharma company. But there was also a desire on the board, and with Bob and me, to turn Domantis into a Genentech. We reached out to private equity groups in New York—would it be possible to raise $200m tomorrow to fund our transition from a medium-sized, quite interesting biotech firm into something much bigger? The answer was no—it was just not possible to raise $200m in private equity for a UK-based biotech firm with no products in the clinic, it was not going to happen.'

Although there were clear benefits from the takeover for the UK life sciences industry, the Domantis sale underlined the financing bind in which UK biotechs now found themselves. With the London Stock Exchange now much less supportive than in the 1990s, emerging UK firms found it more difficult to raise money from the public markets than they might have done a few years earlier. Figure 4.1 shows that firms founded in the 1980s and early 1990s had a higher probability of getting onto the stock market than later entrants. With the waning of stock market support, the number of acquisitions increased sharply in the first decade of the new millennium (Figure 4.2).

---

[76] Although the CAT sale to AstraZeneca remains the second largest acquisition of a UK biotechnology firm after Celltech, its sale returned a little over twice the money invested over its more than 15 years as an independent company. Michael M. Hopkins, Philippa A. Crane, Paul Nightingale, and Charles Baden-Fuller, 'Buying big into biotech: scale, financing and the dynamics of UK biotech, 1980–2009', *Industrial and Corporate Change*, 22, 4 (August 2013) pp. 903–52.

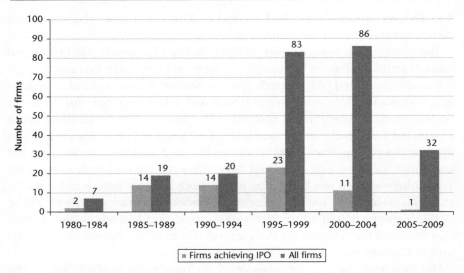

Figure 4.1. Number of UK therapeutic biotech firms founded between 1980 and 2009 (grouped into 5-year cohorts by founding year) showing which of them achieved an IPO
Source: Science Policy Research Unit, University of Sussex.

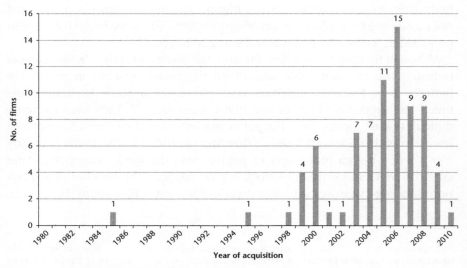

Figure 4.2. Number of UK therapeutic biotech firms acquired, 1980–2010
Source: Science Policy Research Unit, University of Sussex.

The alternative to the IPO was the trade sale, and Domantis was one of several British firms which followed that route. But when was the right time to sell? That was a question on which the views of owners, founders, and managers might diverge.

Selling out Too Soon?

Arrow Therapeutics was founded in 1998 by Ken Powell, who had been head of biology in Wellcome Foundation before joining UCL as a professor in the Virology Research Institute. Arrow, of which Powell became chief executive, was one of several UCL spin-outs which he helped to set up. It specialized in anti-infective medicines, principally anti-virals, which was Powell's area of expertise. Having raised Series A and Series B funding from two groups of venture capitalists (the first round in 2000, the second in 2004), he negotiated a licensing deal with Novartis on one of Arrow's lead drugs, a treatment for Respiratory Syncytial Virus (RSV) lung infections.

This was a useful boost to Arrow's finances, but the company needed to raise more funds to push forward its other products, including a treatment for hepatitis C. During 2006 Powell and his finance director, Ian Garland,[77] considered four options: an IPO on AIM; a reverse merger with a NASDAQ-listed company; a Series C fund raising from existing and new investors; and a trade sale. The first was rejected because the amounts of money raised would be too small. The second was seriously considered—the prospective partner was Inhibitex, which had cash but a weak pipeline—but it was not acceptable to the shareholders. The third was complicated by a difference of view between the first and second group of VCs, with the former pressing for an early exit while the latter were willing to soldier on. In the end the fourth option was chosen, and in 2007 Arrow was sold to AstraZeneca for £76m.

'It was frustrating', Garland said later. 'The takeout was probably not in line with the fundamentals of the business. It would have been different if some of those Series A investors had been evergreen funds or if there had been some other way of allowing them to exit while keeping the company whole. The takeover was a good deal for investors but a fraction of what we might have achieved if we had stayed independent for longer.'[78] Instead Arrow's acquirer, AstraZeneca, took the view, in 2010, that it should exit anti-viral research and close its new subsidiary down, as part of a wider restructuring programme.

---

[77] Garland, trained as an accountant, spent part of his earlier career in the US, working first for Pepsi and then for Medeva, which became part of Celltech. From Celltech he moved to Amarin and then to Arrow.

[78] Interview with Ian Garland, 17 February 2014.

Garland returned to the sector, taking over as chief executive at Acambis, a vaccine firm which had floated as Peptide Therapeutics in 1995, only to find himself in a similar situation to the one that he had faced at Arrow.

Acambis had benefited from the 1990s biotech boom in the UK. In 1999 it was able to raise over £20m in new funds to finance the acquisition of US vaccine developer Oravax, and take a new strategic direction. This brought in a technology which allowed Acambis to develop a smallpox vaccine for the US government's large anti-biological terrorism procurement programme, following the 2001 anthrax attack. But Acambis ran into problems when it failed to win a second US contract. In 2007 the chief executive resigned and Garland was brought in to replace him.

Investors had lost confidence in the business, but Garland and his chairman, Peter Fellner, were convinced that Acambis had a range of technologies that could form the basis for a competitive vaccine company; it also had three valuable partnerships with Sanofi. In 2008 they went on the road to raise more capital, and despite the darkening economic climate they raised £43m through a placing and open offer, the biggest fund-raising in the UK biotech sector for several years.

Meanwhile Sanofi was keen to get its hands on other parts of the Acambis pipeline, but Garland resisted on the grounds that if virtually all its programmes were licensed to one partner Acambis would no longer be regarded as an independent firm. Sanofi then broke the impasse by bidding for the whole company. The price paid was £276m, which the two biggest shareholders, Invesco and Goldman Sachs, were happy to accept.

The outcome was disappointing for Garland, since, in his view, Acambis could have pursued an independent strategy. However, as he said after the deal had gone through, the Sanofi acquisition showed that there was strength in UK biotechnology. 'Perhaps UK investors will take away the message that they can make money in the sector.'[79]

Onset of the World Financial Crisis

Satisfactory though the Sanofi/Acambis deal was for shareholders, it highlighted the continuing decline in the number of sizeable, independent, British-owned biotech firms. There seemed no prospect of reversing that trend as the world financial crisis deepened. In 2008 and 2009 the funding problems facing the biotech sector became even more acute. Promising firms were being bought out and a growing number of cash-hungry companies were

---

[79] *Financial Times*, 25 July 2008.

going into liquidation—a fate only rarely seen in the previous history of UK biotech, acquisitions or mergers being much more common.[80]

In 2008 3i, which had been one of the principal supporters of the sector, announced that it was abandoning investment in start-up firms in order to focus on management buy-outs and late-stage investments. Having suffered heavy write-downs after the technology crash of 2000/2001, 3i had cut back on venture capital so that by 2007 it amounted to less than a tenth of its total portfolio.[81] 'Early-stage has not been an easy place', said Philip Yea, 3i's chief executive—a sentiment shared by Chris Evans, who announced in April 2008 that he was transferring the three funds in Merlin Biosciences to a new company, Excalibur, which would invest in late-stage biotech firms and in other parts of the healthcare sector, including diagnostics and surgical equipment. The environment for early-stage firms, Evans said, was 'fraught with difficulties'.[82]

Not long after 3i's withdrawal Apax Partners sold off most of its early stage investments. Atlas Venture closed down its London office and moved its European operations to Boston. Even firms that still supported early-stage ventures had to adopt a more balanced strategy, with more investment in late-stage firms and in companies which had already gone public. As Stephen Bunting of Abingworth remarked, 'To stay in business we had to innovate away from the old model, we had to change the way venture capital can function. If you just do start-ups you are not going to get the money back fast enough. The financing risk is too great, the holding period too long.'[83]

There was also a recognition within the venture capital community, and among scientists, that the creation of new companies, with all the staffing and other costs that even small firms incur, was not necessarily the most effective way of discovering whether a promising molecule was likely to be commercially viable. One way of keeping costs down was to keep the business 'virtual', employing only a handful of scientists on a full-time basis and out-sourcing clinical development and other functions to outside contractors.

Index Ventures, a venture capital firm which originated in Switzerland and had offices in London and San Francisco as well as Geneva, adopted what it called asset-centric investing. As Francesco de Rubertis, senior partner, explained, a typical biotech firm would have four or five molecules under development, all at different stages of development; there was a tendency to keep spending on all of them, and thus to spread the risk, even if one of them was clearly less promising. 'The company is not ruthless enough in closing down the least promising programmes, and lots of money is wasted on

[80] Hopkins et al., 'Buying big into biotech'.
[81] *Financial Times*, 25 March 2008.   [82] *Financial Times*, 10 April 2008.
[83] Interview with Stephen Bunting, 3 December 2013.

failures.'[84] Under the asset-centric approach separate teams are set up as distinct entities, to focus only on one molecule; the managers are totally committed to that project and have neither the incentive nor the resources to keep investing in it if the early trials show that it is likely to fail. If the molecule showed real promise, it would be sold, probably to a pharmaceutical company; if not, the project would be terminated.

Index launched its asset-centric strategy in 2005 and by 2015 most of its biotech investments took this form. One example was a Cambridge-based firm, XO1, which was developing a synthetic anti-thrombin antibody for the treatment of cardiovascular diseases. It was founded in 2013 by Index, together with Trevor Baglin, a consultant haematologist at Addenbrooke's Hospital and Professor Jim Huntington at the Department of Haematology; the university also had a small shareholding. Led by Richard Mason as chief executive and with David Grainger from Index acting as chief scientist (the two men had worked together at Cambridge Antibody Technology), XO1 was run with only two full-time employees; most functions were out-sourced. It was sold in 2015, 21 months after it had been set up, to Johnson & Johnson for an undisclosed price.

Could the Government Help?

Yet neither the 'virtual' approach nor asset-centric investing could provide a complete answer to the sector's problems. If there was to be a viable biotech sector in the UK it needed a respectable number of firms that stayed independent for long enough to create employment, to make profits and to retain the support of investors, as many US biotechs had done.

Three of the companies that had floated on the main market between 2004 and 2006, Ark, Ardana, and Renovo, suffered disappointments in clinical trials, leading in Ardana's case to the appointment of administrators in 2008. The other two limped on, but the depressed price of their shares made it impossible to raise new funds.[85] AIM-listed firms had not done much better.[86] 'AIM has been a terrible market for biotech and is not attractive for us investors', a partner in 3i remarked. 'Too many companies went on to the

---

[84] Interview with Francesco de Rubertis, 22 March 2013.

[85] None of the three biotech firms that had listed on the main market in 2000 (Table 4.1) had done well. Weston Medical went into liquidation in 2003. Pharmagene was sold to Asterand of the US in 2005, when its market value had fallen to £13m. In the same year Profile Therapeutics was bought by Respironics of the US for £25m.

[86] An exception was Proximagen, backed by the IP Group and Invesco, which floated on AIM in 2005, made several acquisitions, and was sold in 2012 to Upsher Smith of the US for an upfront sum of $347m, rising to $553m if milestones were met.

market too early.'[87] The number of biotech-related firms listed on AIM fell from a peak of 76 at the end of 2007 to just over 40 by 2010.[88]

At the end of 2008 a group of biotech executives and investors led by Sir Chris Evans appealed to the government for help.[89] 'The sector desperately needs large sums in investment', Sir Chris said. 'We are looking down a deep, dark pit. What we do or don't do in 2009 is going to change our industry forever.'[90] He urged the government to invest alongside the private sector in a £500m national consolidation fund, through which professionals would identify the most promising among the hundreds of sub-optimal biotech companies and pool their managements and intellectual property into twenty to thirty enlarged businesses.

A few weeks later Sir David Cooksey's Bioscience Innovation and Growth Team (BIGT) published its second report.[91] The funding climate for biotechnology, the report said, 'is more difficult today than at any time in the short lifetime of the industry'. Sir David and his colleagues frankly recognized that the vision set out in their first report, 5 years earlier, was now even less attainable. Instead of looking forward to the emergence of 'a core of large, profitable, world class companies', the new report set out a more modest goal: to create 'a diverse self-sustaining bioscience sector which supports, on a sustainable basis, high-value-added employment, thereby leading to increased wealth creation and improved health in the UK'.

The BIGT report pointed out that since 2002 the number of public companies with a market capitalization of less than £25m had increased while the number valued at more than £25m had decreased. Companies were no longer able to plan the financing of their operations up to becoming self-sufficient; this had led to trade sales becoming the dominant business model for UK bioscience companies. 'This comes at the expense of creating enough self-sustaining bioscience companies to attract interest within the public markets.' These financing problems had been compounded, it was argued, by greater conservatism on the part of the regulators, making drug development more expensive and protracted in the UK than it was for biotech firms in the US.

Investors, the report said, had become increasingly sceptical about the rate of return likely be achieved by investment in UK biotech. The few British biotechs that had achieved some success as public companies, such as

---

[87] Andrew Franklin, a partner at 3i, quoted in *Financial Times*, 28 May 2007.

[88] Colin Haslam, Nick Tsitsianis, and Pauline Cheadle, 'UK bio-pharma: innovation, re-invention and capital at risk', Institute of Chartered Accountants of Scotland, 2011.

[89] Chris Evans was knighted in 2001.

[90] *Financial Times*, 5 December 2008.

[91] *The Review and Refresh of Bioscience 2015*, Report to Government by the Bioscience Innovation and Growth Team (Department for Business, Enterprise and Regulatory Reform, January 2009).

Cambridge Antibody Technology, had recently been acquired. 'The UK has yet to maintain an independent, successful, large, international bioscience company and, without a demonstration of success through the development of companies to a point where products can generate income and profits, an IPO exit is likely to remain a closed door for bioscience companies.'

The report noted that recent acquisitions of British firms by non-British companies could be regarded as a tribute to the strength of UK research, but 'it also underlines the lack of specialist public market biotechnology investors willing to build a company here and maybe also a dearth of management able to build a strong sustainable company'.

As for remedies, Sir David rejected Chris Evans's suggestion that the government should inject large sums of money into the industry, but he pointed to several areas where the government might provide more assistance. These included extensions to the R & D tax credits to cover expenditure on such items as intellectual property costs and payments to self-employed professionals at senior level; changes in the Enterprise Investment Scheme to cover shareholder-to-shareholder sales; and the provision of tax incentives to encourage large pharmaceutical companies to spin out assets in the UK as the basis for the creation of new firms.

In its response to the report the government pointed out that a widening of the qualifying costs for the R&D tax credits would increase the cost of the scheme with no clear benefit in additional R&D.[92] It was also doubtful about the proposal to extend the applicability of the Enterprise Investment Scheme and Venture Capital Trusts. Some steps had already been taken to simplify and improve these schemes, but a further relaxation of the rules—for example, to allow larger companies to use the schemes—would run foul of the European Commission's guidelines on state aid (which prohibit substantial subsidies that might be anti-competitive or discriminate against foreign firms). Nevertheless, the government accepted the need for a healthy biotech sector and stepped up its efforts to improve the environment for biotech firms, and for the life sciences industry as a whole.

In 2009 the government set up an Office for Life Sciences, chaired by Lord Drayson, the former head of PowderJect, who had succeeded Lord Sainsbury as Science Minister.[93] The task of the Office was to drive 'coordinated, coherent and visible action by government and industry to ensure that the UK is the place of choice for life sciences companies'. It formed part of the Labour

[92] *Government Response to Review and Refresh of Bioscience 2015 Report* (Department for Business, Enterprise and Regulatory Reform, May 2009).

[93] Paul Drayson had been made a working peer in 2004 and entered the Labour government as Minister for Defence Procurement. He held the post of Minister for Science from 2008 until the 2010 election.

government's more activist approach to industrial policy, designed, as Lord Drayson put it, to 'gear up our economy in those sectors showing the greatest potential for long-term growth'.[94]

Biotech was still seen as one of these potentially high-growth sectors. Yet as things stood in 2009 there were few grounds for confidence about the sector's prospects. A commentary contained in *Nature Biotechnology*, an influential industry journal, was damning. 'Despite historic leadership in European biotech', the authors wrote, 'the UK's industry has suffered a near collapse in the past two years and now has little private or public investment and no candidates for world-class companies'.[95]

---

[94] Speech by Lord Drayson at the Academy of Medical Sciences, 11 March 2009.
[95] Graham Smith, Muhammad Safwan Akram, Keith Redpath, and William Bains, 'Wasting cash—the decline of the British biotech sector', *Nature Biotechnology* 27 (2009) pp. 531–7.

5

A Second Chance for UK Biotech

Given that we are on the doorstep of one of the world's great financial centres, there has been a paucity of risk capital for life sciences.... We've never had a Gilead or a Celgene and this is becoming a pressing issue. We've got a couple of pharma companies, which is good, but it's pretty heavy going for them at the moment. Sir John Bell, 29 March 2015.[1]

When the world financial crisis was at its height, most UK biotech firms were focused on survival. With no possibility of raising funds on the public market and only limited access to venture capital, the priority was to cut costs and to conserve cash, and this often involved cancelling or cutting back research programmes. The US biotech sector was also hard hit by the crisis, but investor support began to revive in 2012, leading to a surge in biotech IPOs which continued into 2015. The UK had to wait longer, but the IPO drought on the London Stock Exchange finally ended in 2014, and there was an encouraging increase in venture capital financing to accompany the return of stock market investors. By 2015 the prospects for UK biotech looked better than at any time since the mid-1990s.[2]

More Casualties

In the immediate aftermath of the crisis the flow of bad news continued. A notable casualty was Antisoma, whose record since its foundation in 1988 demonstrates in an extreme form the vagaries of the drug development process. This company was based at the start on technology licensed from the Imperial Cancer Research Fund (later part of Cancer Research UK), which was

[1] Sir John Bell, Regius Professor of Medicine at Oxford University, adviser to the UK government on life sciences, interviewed in *Financial Times*, 29 March 2015.
[2] 'Fundamental strengths of the UK ecosystem: state of the nation 2014', joint report by Ernst & Young and the UK BioIndustry Association (London: 2015).

also an investor. Its principal backer in the early years was the Leventis family, which had close links with the founder and chief scientist, Professor Agamemnon Epenetos; other investors, including 3i, came in later. In 1998 Antisoma listed on EASDAQ, the short-lived European exchange, but switched to London at the end of 1999.

Led from 1998 by Glyn Edwards, who had held senior posts in Celltech and other biotech firms, Antisoma adopted a search and develop strategy, in-licensing drugs (both biologics and small molecule drugs) from academia and other sources. It was described by an analyst as 'bestowed with the strong fundamentals we look for in a product-driven biotechnology company'.[3] In 1999, in the first of a series of partnerships with Big Pharma, Antisoma signed a licensing deal with Abbott of the US, for the joint development of an antibody-based treatment for ovarian cancer.

At the end of 2002 Edwards discontinued the Abbott arrangement and replaced it with a broader agreement with Roche, covering all Antisoma's drugs in clinical development. Roche injected £27m into the company and took a 10 per cent equity stake. The Swiss group might have bought the whole company but, according to a Roche spokesman, 'we want to find creativity and innovation—we don't want to suck that out by absorbing a company'.[4]

In 2004 one of Antisoma'a cancer drugs failed in clinical trials, prompting a 50 per cent fall in the share price. As the *Financial Times* remarked, this was 'an almost perfect illustration of the binary risks inherent in biotech investment—it often feels no different from walking into a casino and sticking everything on red'.[5] Two years later Roche ended collaboration on two other Antisoma drugs.

Undaunted, Edwards negotiated another partnership with Novartis early in 2007. This involved a tumour shrinking treatment which Antisoma had licensed from Cancer Research UK and was thought to have blockbuster potential; it was one of a new class of small molecule drugs that disrupt the flow of blood to tumours (angiogenesis inhibitors). Novartis paid $75m as an upfront fee for the right to develop and market the drug, with a further $25m to come when the drug entered Phase III trials. The agreement was hailed by analysts as a 'great deal' which could generate as much as $890m in milestone payments if all went well.[6] Once again the outcome was disappointment. Phase III trials were abandoned early in 2010 when, to the surprise of both companies, they showed that the drug produced no significant improvement over existing therapies. The share price collapsed, and although Antisoma had other drugs under development it could not raise the funds to take them through clinical trials. In 2011 all drug programmes were halted; the company

[3] S. G. Cowen, *Equity Research*, January 2002.   [4] *Financial Times*, 9 December 2002.
[5] *Financial Times*, 27 April 2004.   [6] *Financial Times*, 20 April 2007.

moved from the main market to AIM and reclassified itself as an investment firm, with no further involvement in biotech.

This was a sad end for a firm that had negotiated several potentially rewarding partnerships with blue-chip pharmaceutical companies. As Edwards said later, 'We wanted to take products as far as we could before partnering—you get more value that way—but if you partner and the drug fails you take a big hit and you don't get the revenues. I am sceptical of the view that partnering reduces the risk.'[7]

The demise of Antisoma could not be attributed to lack of finance, or to poor management. Glyn Edwards and his chief operating officer, Ursula Ney (who had also worked for Celltech in its early days), went on to hold senior posts in other biotech firms.[8] But, coming as it did on top of other failures, the collapse of Antisoma confirmed investors in their distrust of the biotech sector.[9]

Of the four firms which had listed on the LSE main market between 2004 and 2006, only Harry Stratford's ProStrakan generated a positive return on investment. In 2011 it was sold for £300m to Kyowa Hakka Kirin, which was a licensee for one of ProStrakan's drugs. Ardana went into administration in 2008. Ark suffered a crippling blow at the end of 2009 when its lead drug, Cerepro, a gene-based therapy for the treatment of brain tumours, failed to win approval from the European regulators; it used its remaining resources to purchase a veterinary business. Renovo's anti-scarring treatment, Juvista, failed in Phase III trials early in 2011. Once regarded as a rising star, Renovo laid off most of its staff and turned itself into a cash shell.

Venture Capitalists Look for Quicker Exits

Between 2006 and 2014 there were no biotech flotations on the LSE's main market. Stockbrokers who had raised capital for biotech firms pulled out of the sector or switched to diagnostics and medical devices; Piper Jaffray, a leading adviser to UK healthcare companies, closed down its European equities business in 2010.[10] London-based investment trusts which specialized in biotech, such as the Biotech Growth Trust, concentrated almost entirely on US companies.

---

[7] Interview with Glyn Edwards, 30 September 2013.

[8] Edwards was later appointed chief executive of Summit (formerly VASTox), while Ursula Ney became chief executive of Genkyotex, a Swiss biotech firm.

[9] Another casualty was Alizyme, which went into administration in 2009. After moving from AIM to the main market in 2000, this company developed an anti-obesity drug which was thought to have blockbuster potential. It was licensed to Takeda in Japan, but Alizyme was unable to find a partner in the US. This setback, together with clinical trial failures for other drugs, severely weakened the company's finances. *Financial Times*, 30 June 2009.

[10] *Financial Times*, 10 November 2010.

Some venture capitalists were still active in the sector, but they were looking for quicker exits. Advent was the lead investor in Respivert, founded in 2007 to work on treatments for Chronic Obstructive Pulmonary Disease and other respiratory illnesses; two of the founders had previously worked in the same field at GSK. When the company was sold 3 years later to Johnson & Johnson, Advent described the deal as a validation of its 'strategic approach' to portfolio building, which was designed to produce earlier returns than those typically associated with venture funds. An Advent partner, Shahzad Malik, said the firm had moved away from long holding periods and multiple funding rounds. 'We don't want to spend $100m on a business and rely on public markets to bail out investors. It's a broken model. We are looking for pharma companies to buy early, even if we run the risk of a lower margin.'[11]

Another Advent-backed company, Thiakis, was founded in 2004 to exploit technology developed by Professor Stephen Bloom at Imperial College for the treatment of obesity and diabetes; the co-founder and chief executive was John Burt, whose previous experience had included a spell with GSK, working mainly on acquiring technology from small biotech firms. The company was run on a 'virtual' basis, with only a handful of permanent employees and extensive use of outsourcing. By 2008 some promising efficacy data had been assembled, and the firm had to choose between doing a Series B round and selling to Big Pharma. 'We had kept in close touch with pharmaceutical companies', Burt said, 'and we always expected to sell. Advent wanted an exit and a Series B round would have been very difficult at that time.'[12]

Thiakis was bought by Wyeth of the US for an upfront payment of £20m, with another £80m payable if milestones were reached. Wyeth was taken over by Pfizer soon after the Thiakis deal had gone through, but Pfizer subsequently decided to withdraw from obesity drugs and discontinued the Thiakis programme; no further payments after the initial £20m were made. The intellectual property was returned to Imperial Innovations.

An early exit was more likely if the firm was already well advanced in its drug programmes when it obtained venture capital support. More opportunities of this kind became available as the big pharmaceutical companies, under pressure from investors to get a better return from their investment in research, withdrew from areas such as neurological and psychiatric diseases where they had difficulty developing innovative medicines. In 2010 GSK was divesting assets and laying off staff, including the closure of what had been a substantial R&D centre at Harlow, Essex. Convergence Pharmaceuticals acquired from GSK a set of small molecule drug candidates for the treatment of neuropathic

---

[11] Quoted in *realdeals*, 30 June 2010.   [12] Interview with John Burt, 6 August 2014.

pain in 2010; it was sold in 2015 to Biogen Idec in a deal potentially worth around £400m, dependent on future development milestones being met.

The stream of trade sales, together with the demise of Antisoma and others, raised fears, as one newspaper headline put in, that 'Britain's biotech stars are fading away'. Biotech had been seen as a sector where Britain could lead the world; now, according to the article, it seemed to be another example of how Britain squandered its innovations in science and engineering.[13] This prompted a riposte from Kate Bingham of SV Life Sciences, which had been an investor in several of the firms recently sold to Big Pharma.

Bingham insisted that trade sales, and the absence of IPOs, should not be seen as a sign of weakness. Most pharmaceutical companies had moved to some extent away from relying on in-house R&D in favour of buying in new development stage products from outside. 'The reason why the UK public biotech market is shrinking', she said, 'is thus a simple one. Successful early stage biotech companies are being increasingly bought directly by big pharmaceutical companies. These are able to devote substantial funds and expertise to developing such life-changing new drugs, something that the older small public biotech companies cannot do because of a shortage of cash. Innovation in the UK is thriving and healthy. Patients will continue to benefit from novel therapies discovered in UK biotech companies, even if these companies never tap the public markets.'[14]

Bingham's argument was that in most cases a novel drug, once the biotech firm had demonstrated its therapeutic value, was likely to be developed more quickly and more effectively inside Big Pharma. The Thiakis case showed that this outcome could not be guaranteed. The drug might fail in clinical trials, or the acquiring company might change its mind and decide not to develop the asset it had bought.

Others believed that it would be better for the health of UK biotech if some British firms stayed independent for longer, developing their business to the point where they could come to the stock market and continue to grow as public companies, as many American firms had done.[15] The 2002 BIGT report had argued that the UK needed beacons to encourage investors and entrepreneurs to put money into biotechnology.

---

[13] Julia Kollewe, 'Britain's biotech stars fade away', *The Guardian*, 29 August 2011.

[14] Kate Bingham, Letter to *The Guardian*, 1 September 2011.

[15] An analysis by one of the authors of 127 UK therapeutic biotech firms founded in the 1980s and 1990s showed that those which had received stock market funding alone, or in combination with venture capital funding were more likely to have produced a drug that passed Phase II trials than firms which had received only venture capital funding. This is due to the greater funds raised and longer time as independent entities that these firms had compared to biotech firms funded in other ways. Michael Hopkins, Philippa A. Crane, Paul Nightingale, and Charles Baden-Fuller, 'Buying Big into Biotech: scale, financing and the dynamics of UK biotech, 1980–2009', *Industrial and Corporate Change* 22, 4 (August 2013) pp. 903–52.

Still Looking for Beacons

The beacons sought by the BIGT were nowhere to be seen. As the UK biotech sector came towards the end of its third decade, there was still a dearth of outstanding successes, whether in terms of consistently profitable firms or high selling innovative drugs. Table 5.1 sets out the novelty and molecule type of the first 100 launched drugs that the UK biotech sector played a role in bringing to market (counting all molecules that the UK firm had a stake in, however small, and whether they were involved in marketing or not). The table distinguishes between lead molecules (the first in their class brought to market), molecules that follow established strategies (follow-on drugs), and new formulations (including novel dosage and drug delivery strategies for known drugs).

Of the eight lead molecule biologicals brought to market, three were monoclonal antibodies—Mylotarg (launched in 2000) Campath (2001), and Humira (2002). A fourth monoclonal antibody, Cimzia (2008) was technically a new formulation. Each was ultimately brought to market by a foreign owned firm. The only recombinant proteins that UK firms brought to market were those that Shire had acquired from its acquisition of TKT in the US. Most UK biologics, including a number of vaccines, followed more established therapeutic strategies.

Of the nine lead chemical drugs brought to market most were niche products, such as Oxford GlycoSciences' Zavesca, although Shire had a hand in three that it had bought in for development, as well as another three drugs that followed established therapeutic strategies. The bulk of the UK biotech sector's output (41 out of 100 approved drugs) consisted of new formulations of existing drugs, a low risk, low margin strategy. In natural products, GW Pharma's Sativex (nabiximols) proved that the oldest pharmaceutical tradition of developing drugs from botanical extracts could still pay off for some, although others following this strategy had less success.

The fact that only Shire had managed to become a big commercial success, and that it had done so without relying on UK-originated drugs, left the biotech sector and its investors still searching for inspirational firms that could bring

Table 5.1. The first 100 marketed drugs from UK biotech firms

| | Novelty | | | |
	Lead compound	Established Strategy	New Formulation	Unknown
Biological	8	12	1	4
Chemical	9	12	41	3
Natural Product	3	0	7	0

Source: Science Policy Research Unit, University of Sussex (based on data provided by Pharmaprojects).

blockbuster drugs developed in British laboratories to the market. Would the newer firms, drawing on the experience of the first generation, do better?

Mission Therapeutics, founded in Cambridge in 2011, was the second company formed by Steve Jackson, professor of biochemistry at the Gurdon Institute.[16] Jackson, who had spent part of his earlier career in the US, had set up KuDOS Pharmaceuticals in 1997, with seed funding from Cancer Research Campaign, to develop small molecule drugs that block DNA repair in cancer cells. In 2005 the lead molecules were about to enter clinical trials when KuDOS was bought by one of its partners, AstraZeneca. Although Jackson would have preferred KuDOS to have stayed independent for longer, he regarded the outcome as a success. 'My idea is to make a difference', Jackson said. 'One way or another companies like KuDOS will probably end up in Big Pharma and that is OK, all the more so if it is a UK-based pharma company.'[17]

Mission's goal was similar to that of KuDOS, to develop small molecule drugs primarily in oncology. Mission was backed at the start by Cancer Research Technology, and it attracted a group of venture capital firms which included Sofinnova from France, Imperial Innovations and two corporate VCs, GSK's SR One and Roche Venture Fund; a third corporate VC, Pfizer Venture Investments, invested in the Series B financing round at the end of 2013.

'The VCs were very enthusiastic that we could win big and beat the competition', Jackson said, 'They shared my vision of the company—to think big as the Americans do, not just developing a couple of products that can be sold to Big Pharma. That sort of attitude is now more prevalent in the Cambridge cluster—it is much less parochial than in the past.'

The most famous of Cambridge's scientist-entrepreneurs was Sir Greg Winter, co-founder of Cambridge Antibody Technology and Domantis.[18] In 2009 he set up his third company, Bicycle Therapeutics, focused on a new class of drug called bicycle peptide drug conjugates; these are smaller than antibodies and have the advantage of penetrating the tissue to attack tumours, especially solid tumours, more quickly.

One of Winter's former post-doctoral students, Regina Hodits, who was in charge of Atlas Venture's biotech investments in Europe, had asked him to keep her informed of any new venture he might be considering. She had been

---

[16] The Gurdon Institute had been set up in 1989 as the Wellcome Trust and Cancer Research UK Institute, part of Cambridge University's School of Biological Sciences. It was renamed the Gurdon Institute in 2004 in honour of Professor Sir John Gurdon, one of the founder members; Professor Gurdon was awarded the Nobel Prize for Physiology or Medicine in 2012.

[17] Interview with Steve Jackson, 18 June 2013. In 2014 a treatment for ovarian cancer which had been developed by KuDOS and taken on by AstraZeneca, olaparib (Lynparza), was approved by the FDA.

[18] Winter was knighted in 2004. He was appointed Master of Trinity College, Cambridge in 2002.

looking for investment opportunities in the area in which Winter and his colleague, Christian Heinis, were working, and she took the lead in securing the relevant intellectual property for Bicycle and helping to build the management team. Atlas was one of the founding investors, along with SV Life Sciences and two corporate VCs, Novartis Venture Fund and SR One; a third corporate VC, Astellas Venture Management from Japan, came in later. Thus three of the five investors were evergreen funds, less concerned than traditional VCs about the timing of exit.

Hodits said later: 'It certainly helped that investors had made money with Sir Greg's earlier companies, CAT and Domantis. That helped to open doors, but in the end it comes down to the value that that can be seen in the platform and the team that is involved in the company.'[19]

The fact that both Mission and Bicycle were supported by non-British as well as British investors showed that, despite earlier disappointments, British science and British scientist-entrepreneurs were highly regarded around the world.[20] The same was true of Heptares, created in 2007 to exploit discoveries made by Richard Henderson and Chris Tate at the MRC's Laboratory of Molecular Biology. The technology was aimed at a class of drug targets known as G-coupled protein receptors, and had the potential to produce novel treatments for Alzheimer's disease and schizophrenia.

Seed funding came from MVM, which originally had been created by the Medical Research Council and still maintained close links with it, but the Series A round in 2009 was led by a US venture capital firm, Clarus Ventures. Michael Steinmetz, managing director of Clarus, who had formerly been head of biology at Roche, joined the Heptares board. Heptares had an experienced chairman in John Berriman (ex-Celltech and ex-Abingworth), while the chief executive and co-founder, Malcolm Weir, had experience in Big Pharma as well as running a start-up firm.[21]

These three were among several promising firms that had the potential to grow into substantial businesses. But there was concern in government—and this applied as much to the Conservative–Liberal Democrat coalition which took office in 2010 as to its Labour predecessor—that too many firms of this kind were being sold too early. Building biotech firms with a view to an early sale was not necessarily the best way of maximizing the value of British science.

---

[19] Email communication from Regina Hodits, 4 August 2015.

[20] These firms were also able to attract top managers from Big Pharma. In 2015 Kevin Lee, who had been in charge of rare disease research at Pfizer, was appointed chief executive of Bicycle Therapeutics.

[21] Weir had headed the molecular sciences division of GlaxoWellcome before starting Inpharmatica, a specialist in structural bioinformatics, in 2000; Inpharmatica was sold to Galapagos of Belgium in 2006.

121

There was also anxiety, highlighted by Pfizer's decision in 2011 to close its Sandwich laboratory, about moves by several Big Pharma companies to cut back their research operations in Britain. The life sciences industry as a whole would be healthier, according to this view, if it contained a tier of solidly profitable mid-cap firms, filling the gap between GlaxoSmithKline and Astra-Zeneca at the top and the numerous small biotechs down below. That was more likely to happen if biotechs had access to alternative sources of capital, not constrained by exit deadlines.

New Sources of Finance

In 2012, partly in response to these concerns, the Coalition government launched the Biomedical Catalyst fund, a £180m fund run jointly by the Medical Research Council and the Technology Strategy Board (later renamed Innovate UK).[22] Its purpose was to speed up the development of health-related innovation through grants ranging from £150,000 for feasibility studies up to £2.4m for projects entering clinical trials. Zahid Latif, head of healthcare at the Technology Strategy Board, explained the rationale. 'Our intention is to help senior management in companies think more constructively about sources of finance. Instead of scrambling for money from any source and probably thinking of an early exit, we want them to take a longer-term view and build the company to a bigger size, or potentially take on greater risk/greater reward type projects.'[23]

The emphasis on projects rather than companies was part of the thinking behind the establishment of a new fund set up by Cancer Research Technology, the development and commercialization arm of Cancer Research UK.[24] The purpose of the CRT Pioneer Fund, launched in 2012, was to speed up the process of drug discovery and development from early lead optimization to the end of Phase I trials. The aim was to create a new path from research to the clinic, and then to license the drug candidate to an established company for onward development into Phase II and beyond.

Keith Blundy, chief executive of Cancer Research Technology, believed that, while some investments by the Pioneer Fund might lead to the creation of a new firm, licensing in most cases was a better option. 'Our thinking is—what is the best way of developing the technology? There are already lots of

---

[22] The Technology Strategy Board had been set up in 2004 to take responsibility for the industrial support schemes run by the Department of Trade and Industry. In 2007 it was made an independent non-departmental agency (see Chapter 8).

[23] Interview with Zahid Latif, 3 February 2014.

[24] Cancer Research UK was formed in 2002 by a merger between Cancer Research Campaign and the Imperial Cancer Research Fund.

companies out there—why create more with all the costs of infrastructure, overheads and so on? What matters is to get drug development quickly into the hands of a partner.'[25] Cancer Research Technology does continue to create start-up firms—Steve Jackson's Mission was one—but only where the firm concerned has the potential to generate multiple assets, rather than a single project.

The biggest charitable funder of biomedical research in the UK, the Wellcome Trust, was also considering new funding initiatives. The bulk of the Trust's income was spent on academic research in universities, but as part of its diversified investment portfolio it also invested in biotech firms, either as a limited partner in venture capital funds like Abingworth or directly as an investor in start-up firms, usually in partnership with venture capitalists.

The first sign of a new approach came in 2009 when the Trust supported a new monoclonal antibody firm, Kymab, in Cambridge. Spun out from the Sanger Institute (itself supported by Wellcome), Kymab was set up to exploit a new way of generating monoclonal antibodies, based on the Kymouse transgenic mouse platform which had been developed by Professor Allan Bradley.[26] The advantage of the platform was that it should produce a broader repertoire of selective and potent human monoclonal antibodies, providing a range of potential candidates for drug development.

The expectation at the start was that Wellcome would bring in venture capital firms as co-investors. But the Trust decided that, at least for the first phase, Wellcome would be the sole investor; the initial investment was £20m. Andrew Sandham, an experienced biotech manager who was appointed Kymab's chairman and chief executive to get the company started, explained the Trust's thinking.[27] 'If you want to build a platform and really maximise its value you have to back a number of projects and partner quite broadly—you have to think 10–15 years, not 5–8 years.'[28]

If venture capital firms came in later, they would have to accept the same long-term approach as Wellcome. In 2014 Wellcome brought in as a co-investor the Bill and Melinda Gates Foundation, which invested $20m in Kymab, alongside an additional $20m from Wellcome Trust. During 2015 this round was expanded further to $90m with investment from Woodford

---

[25] Interview with Keith Blundy, 12 May 2013.

[26] Professor Bradley had spent part of his career in the US, working at the Howard Hughes Medical Institute. He had been involved in creating several biotech firms in the US, including Lexicon Genetics.

[27] In 2013 Sandham was succeeded as chairman by David Chiswell, formerly chief executive of Cambridge Antibody Technology, and as chief executive by Christian Groendal, who had previously worked for Zealand Pharma and Novo Nordisk.

[28] Interview with Andrew Sandham, 17 June 2013.

Patient Capital Trust and Malin Corporation, both of which had a long-term investment strategy.[29]

Wellcome's next step, in 2012, was to create a £200m healthcare investment fund, Syncona Partners, which would seek to build healthcare businesses on a long-term basis. Syncona got under way at the start of 2013, with Martin Murphy, formerly with MVM, as chief executive, and Nigel Keen, who had long experience in the healthcare business, as chairman.[30] One of its first investments was NightstaRx, based on the development by Professor Robert MacLaren at Oxford's Nuffield Laboratory of Ophthalmology of a gene therapy for an inherited form of progressive blindness known as choroideremia.[31] By 2015 Syncona had a portfolio of five early-stage healthcare firms.

The Syncona initiative was welcomed by biotech firms, many of which were looking for ways of reducing their reliance on venture capital.[32] The funding environment was becoming more diverse, with several new players entering the field. In 2013 Cambridge University set up a new investment fund, Cambridge Innovation Capital, backed by outside investors including Arm Holdings, the semiconductor company, as well as Invesco, Lansdowne Partners, and IP Group.[33] The fund planned to invest with a time horizon of a decade or more, longer than that of a traditional venture capital firm, in the hope that more firms would stay independent for longer and achieve a higher valuation.

Business Angels: The Avidex Story

An increasingly important source of finance was the business angel community. Two Oxford-based investors, Ian Laing and Nick Cross, whose early investments were described in Chapter 3, were involved in a remarkable comeback story which did not depend on venture capital.

This story begins in 1999 with the creation of Avidex, based on the use of T-cell receptor technology to create bispecific immunotherapies that can hunt down cancer cells and make them visible to the human immune system; the company was located in Milton Park, the home of a growing number of

[29] The Woodford Patient Capital Trust was set up by Neil Woodford after he had left Invesco to establish his own investment business (see the section below, 'British Institutional Investors').

[30] Nigel Keen had been a co-founder of Biocompatibles in 1985, and later chairman of Axis-Shield, Oxford Instruments, Laird, and several other companies.

[31] In November 2015 the $35m Series B funding round for NightstaRx was led by New Enterprise Associates of the US, another indication of the growing interest from US venture capitalists in UK biotech.

[32] William Bains, Stella Wooder, and David Ricardo Munoz Guzman, 'Funding biotech start-ups in a post-VC world', *Journal of Commercial Biotechnology* 20, 1 (January 2014).

[33] The chief executive of Cambridge Innovation Capital was Peter Keen, who had worked with Chris Evans at Enzymatix and was one of the biotech sector's most experienced managers and investors.

biotech firms. The founding scientist was Bent Jakobsen from Denmark, who was working at the Institute of Molecular Medicine; he had been invited to Oxford by Professor John Bell. The chief executive (and a founding investor) was James Noble, a veteran of the biotech scene; he had been finance director of British Biotech and a non-executive director in several other firms.

Avidex raised £10m at the end of 2000, principally from Advent, supported by an Oxford-based venture capital trust and a group of private investors including Cross and Laing; Cross joined the board at that time. A further £11.6m was raised in 2002, with Advent again taking the lead. At this stage Jakobsen was preparing for *in vivo* trials, but over the next 2 years it became clear that additional finance would be needed. With Advent no longer able to provide more funding, Noble was faced with the choice of closing down the business or selling it.

In 2006 he found a buyer in Medigene, a German company. Two years later, after a change of strategy in Medigene, Nick Cross, working with Jakobsen and Noble and supported by Laing and George Robinson, a hedge fund manager, bought back the Avidex technology and created two new firms, Adaptimmune and Immunocore; they shared physical and scientific resources and were based in the same building at Milton Park. Other private investors were invited in; they had the same shareholdings in both companies.

Both these firms are pioneers in immuno-oncology, using the patient's own immune system to combat cancer. Both use T-cell receptor technology, but they have different therapeutic approaches. Adaptimmune had developed a process which uses an individual patient's T-cells and re-engineers them so that, when re-administered to that patient, they bind with and destroy the cancer cells. Immunocore had a more generic process, using synthetic T-cell receptors with very high affinity that could bind to cancer cells and mark them for destruction by the patient's unmodified T-cells.

A critical challenge for both firms was to advance their technologies to the point where they would attract the interest of Big Pharma. In 2013 and 2014 Immunocore signed four partnership agreements—with Genentech, GSK, AstraZeneca, and Eli Lilly. As Noble explained, 'we have proved that there are lots of targets which are suitable for T-cell receptors and which antibodies cannot reach. So suddenly Big Pharma has come across a company with 20-plus targets which have been validated scientifically, and we are the only people that can do it—there is no American company in the field.'[34]

The four partnerships brought in sizeable up-front payments, with the promise of more to come as the partnered drug candidates progressed

---

[34] Interview with James Noble, 3 October 2013.

through clinical trials. All this had been achieved without venture capital support, and this was a deliberate choice. Nick Cross, who served as chairman of Immunocore and Adaptimmune until 2014, said: 'We could not run the business properly with the constraints which VCs impose. It is a long-term play and you cannot do that with a fund that has to be in and out in five plus five years.'[35]

The partnerships gave Immunocore the financial capacity to pursue the development of its principal unpartnered drug candidate, a treatment for melanoma. By early 2015 this drug was in Phase IIa trials, and the company was considering how best to finance the more expensive stages of clinical development. A listing on NASDAQ might have been feasible, but in July Immunocore raised $320m in a private financing round, one of the biggest venture rounds ever achieved by a European biotech firm.[36]

Meanwhile Adaptimmune, after struggling for some years to interest Big Pharma in its unconventional technology, negotiated a collaboration agreement with GlaxoSmithKline in June 2014, covering the development and commercialization of its lead cancer programme; the deal could generate payments of over $350m over a 7-year period. By now Adaptimmune had established itself as a promising contender in the cell therapy area of immuno-oncology, which was attracting intense interest from US investors. One of the biggest NASDAQ IPOs in 2014 was that of Juno Therapeutics, an early-stage US firm operating in the same field as Adaptimmune, although with a different technology; Juno raised $265m in its IPO and achieved a market capitalization of over $2bn.

There was clearly an opportunity for Adaptimmune to tap into this investor enthusiasm, and several US venture capitalists were willing to support it. The first step, in September, 2014, was a $104m financing round led by New Enterprise Associates of the US, and this was followed in May 2015 by an IPO on NASDAQ, which raised $176m and valued the company at around $1.3bn.

Immunocore, with the additional financial resources at its disposal, was under no pressure to go for an IPO, but when it did so, it might choose to list on NASDAQ, as its sister company had done, rather than in London. While private capital was flowing into UK biotech at an encouraging rate, public market investors in the UK still had little appetite for biotech IPOs.

---

[35] Interview with Neil Cross, 22 October 2013.

[36] One of the participants in this financing round was Neil Woodford, through Woodford Investment Management. This was the investment firm which he had set up after leaving Invesco in 2013. Eli Lilly, one of Immunocore's Big Pharma partners, also contributed to the financing.

London Stock Exchange versus NASDAQ

As long as British institutional investors were reluctant to support British biotech firms, NASDAQ would continue to be a more attractive venue for IPOs than the London Stock Exchange. But, as earlier experience had shown, moving to NASDAQ was not a simple matter. The ground had to be prepared carefully. US investors had to be convinced, through a programme of presentations that might last for a year or more, that the British firm's scientific assets had real potential as the basis for bringing high-value drugs to the market, and that it had an experienced team capable of handling all the steps necessary to get through the regulatory process.

One firm that met these conditions was GW Pharmaceuticals, which listed on NASDAQ in 2013. The chairman was Geoffrey Guy, who had trained as a medical doctor but had spent most of his career as an entrepreneur in the life sciences.[37] He had founded GW, a specialist in cannabis-based drugs, in 1998, and it went public on AIM in 2001. Over the next few years GW shares had what Justin Gover, chief executive, describes as a roller-coaster ride on AIM, while making steady progress with its lead drug, Sativex, a treatment for multiple sclerosis.[38]

In 2009 this drug showed positive results in Phase III trials. 'This was our big breakthrough', says Gover, 'but it hardly registered with our UK investors. We saw this event as a validation of our science and capabilities, allowing us to move forward in maximising the potential, not just of that product, but of our pipeline, whereas our investors saw it as the end of the journey, they were focused on near-term profitability. I saw that there was a difference between what we wanted to be as a company and what our investors expected of us.'[39]

At this point Gover began to cultivate US investors, who had a deeper understanding of the science and a greater willingness to provide the capital for the next stage of GW's development. In 2013 Guy and Gover, taking advantage of the IPO revival in the US, listed GW shares on NASDAQ.

In the short term the share price did not appreciate much, but in the autumn of that year GW reported positive news about Epidiolex (cannabidiol), a treatment for intractable epilepsy which was about to enter Phase II trials. Investor interest in this drug increased after a physician-led study

---

[37] Before setting up GW, Guy had been part of the first wave of UK biotech start-ups in the 1980s. He founded Ethical Holdings, a drug delivery firm, in 1985, and took it to NASDAQ in 1993. Justin Gover joined Guy at Ethical in 1995 and they considered a London listing in 1997, but market sentiment at the time led them to decide against it. Ethical Holdings, later renamed Amarin, remained a US-listed company.

[38] GW listed on AIM in 2001 at a price per share of 185p. The share price hit a high of 261p in May 2003, reached a low of 26p at the end of 2008, and stood at 64p in May 2013, the day of the NASDAQ listing.

[39] Interview with Justin Gover, 15 June 2015.

showed that children with treatment-resistant epilepsy who had been given the drug had significantly fewer seizures, and some of them had become seizure-free. The share price began to rise sharply and in January 2014 GW was able to raise $100m for the development of the epilepsy drug.

As the Epidiolex programme continued to make progress, GW raised a further $160m in June 2015 and $200m in May 2015. Although GW has retained its listing on AIM and its principal R&D base is in the UK, 75 per cent of its shareholders are in the US. When GW listed on NASDAQ its market capitalization was $150m; in June 2015 it was $2.4bn.

While GW was fortunate in its timing—the coincidence of the IPO boom on NASDAQ and the emergence of Epidiolex as a potential winner—Gover has no doubt that GW could not have raised the capital that it needed if it had remained listed only in London. As he told his shareholders, 'The US has an enormous breadth and depth of investors who are extremely well educated on biotech and manage funds dedicated to the sector. In the UK we were relying for the most part on generalist small cap investors who understandably feel less comfortable with the scientific risks associated with biotech investing. We have an audience now that understands what we are doing in a way that was not possible in the UK.'[40]

GW's example was followed in 2015 by Summit, another long-established UK biotech. It had been founded by Oxford academics as VASTox in 2003 and was focused at the start on chemical genomics; its scientific founders were Steve Davies, professor of organic chemistry and his then wife Kay Davies, who was head of the department of human anatomy and genetics.[41] The principal shareholder was the IP Group, which, as noted in Chapter 4, had a special relationship with the university's chemistry department; VASTox was listed on AIM in 2004, and changed its name to Summit in 2008.

Summit's most promising drug candidate was a treatment for Duchenne Muscular Dystrophy (DMD), a disease for which there was no known cure. It is caused by gene mutations linked to the protein necessary for muscle strength, and affects some 15,000–20,000 boys in the US. Some of the funding for Summit's clinical trials came from muscular dystrophy charities in the US. After Glyn Edwards, formerly with Antisoma, was appointed chief executive in 2012, he concentrated most of the firm's development effort on DMD, drawing on scientific support from Kay Davies at Oxford.

DMD was a therapeutic area that was attracting considerable investor interest in the US. Two of the leading US firms in the field, Sarepta and PTC Therapeutics, were listed on NASDAQ with high valuations, and a Dutch

---

[40] *Financial Times*, 28 February 2014.

[41] In 1992 Steve Davies had founded Oxford Asymmetry, one of the earliest Oxford University spin-offs. It was sold in 2000 to Evotec of Germany for £316m.

firm specializing in DMD, Prosensa, joined them in 2013.[42] It was logical for Summit to want to be seen as part of this peer group. Its shares were listed on NASDAQ in February 2015; it raised $34m (slightly less than it had hoped) to fund further clinical trials. Glyn Edwards said that by joining NASDAQ Summit would gain access to a large number of specialist healthcare investors, and allow its investors to benefit from the additional liquidity that a dual listing would generate.

As earlier experience had shown, a listing on NASDAQ was no guarantee of instant success. For example, Cyclacel's share price had moved erratically after it transferred to the US in 2006, but that did not prevent it from raising over $100m through NASDAQ and pushing forward the development of its cell cycle control technology for the treatment of cancer; in mid-2015 its lead compound, targeted at acute myeloid leukaemia, was in Phase III trials.[43]

British firms that had no presence in the US and no visibility among US investors had no alternative but to go for a listing in London, either on the LSE's main market or on AIM. The issue for them was whether they could drum up sufficient support among UK-based investors to make an IPO in London feasible.

British Institutional Investors

Many investing institutions had lost faith in UK biotech after the setbacks of the late 1990s and early 2000s. The partial recovery of 2004–06 was soon overtaken by the financial crisis, bringing still more bad news from the biotech sector. Those publicly listed firms that attracted institutional support were for the most part ones which had drugs that were already on the market or nearing regulatory approval.

The outstanding example was BTG, which continued to make good progress towards the goal set by Louise Makin of achieving annual sales in excess of $1bn. Most of the growth was coming from interventional medicine, which included Varithena (polidocanal injectable foam), a newly approved treatment for varicose veins, and two products arising from acquisitions made in the US—a radiation treatment for liver tumours and an advanced treatment for blood clots. In another US acquisition at the end of 2014 BTG bought PneumRx, whose lead product was a treatment for emphysema. BTG was

---

[42] In 2014 Prosensa was bought by BioMarin, a larger US biotech firm, for up to $860m. BioMarin had previously had a partnership with Summit but that was discontinued in 2010 after disappointing clinical trials.

[43] Cyclacel retained its research facility in Dundee. In 2015 it had a full-time staff of twelve in Dundee, with much of the R&D out-sourced to contract research organizations. Cyclacel has received two substantial grants from the UK government's Biomedical Catalyst fund.

solidly profitable and had a proven strategy which investors could understand. While Invesco was its largest investor, the share register included several other mainstream institutions, including M & G, and insurers such as Aviva, Standard Life, and Legal & General.

Several of these investors also supported Vernalis, which had had a more chequered record than BTG but was moving in the same direction; it was assembling a portfolio of marketed drugs and limiting its commitment to early-stage research. Vernalis had run into financial problems in 2007 when its application to use its anti-migraine drug, Frova, for menstrual migraine was turned down by the FDA. Ian Garland, who after his success at Acambis had established a reputation for turning round troubled companies, was brought in as chief executive.

Garland's strategy was to build up the company's cash generating businesses and to invest only selectively in research. In 2012 he negotiated a partnership with a US company, Tris, which was developing a range of slow-acting cough and cold medicines, based on a proprietary technology. The first product of the partnership was a new cough medicine approved by the FDA in May 2015. Garland described the approval as 'a very significant moment in the evolution of Vernalis to a commercial stage specialty pharmaceutical company'.[44] While Garland retained a drug discovery team in Cambridge (focused mainly on partnerships with Big Pharma), he was steering Vernalis away from its biotechnology roots.

Generalist institutional investors were less likely to invest in high-risk, early-stage firms, where Neil Woodford at Invesco continued to plough a lonely furrow. Some of his biotech investments were made through Imperial Innovations and IP Group, but Woodford also invested directly in firms that had no connection with these two organizations. E-Therapeutics was a pioneer in network pharmacology, the use of network analysis to determine the set of proteins most critical to any disease. It came to AIM in 2007, and began to prepare clinical trials for its lead drug, which targeted the ability of cancer cells to resist their own self-destruct mechanisms. By 2010 it was running out of cash. With biotech shares in the doldrums the funding outlook looked bleak. A programme of presentations to investors that started towards the end of that year was unexpectedly successful, bringing in Invesco as the principal supporter of a £17m capital issue. Invesco subsequently increased its shareholding to just under 50 per cent. 'They have got great science, a brilliant team, and a charismatic leader in Malcolm Young', Woodford said. 'With our backing they can accelerate work on their lead drug for brain cancer.'[45]

---

[44] Vernalis press release, 1 May 2015.
[45] Interview with Neil Woodford, 6 May 2013.

Woodford also gave his imprimatur to an older firm which had suffered setbacks in its early days, the stem cells specialist, ReNeuron. Seed funded by Merlin in 1997, it was listed on AIM in 2000, delisted in 2003, and brought back to the stock market in 2005. Over the next few years it took its lead compound, a treatment for disabled stroke patients, into clinical trials, and the prospects for the business looked sufficiently promising to permit a £33m refinancing in 2013. Invesco participated, as did Abingworth and ReNeuron's original backer, Chris Evans, through his recently established Arthurian Life Sciences Fund.[46] Helped by Evans's Welsh connections, the transaction was supported by the Welsh government through the Wales Life Sciences Investment Fund. ReNeuron's headquarters were moved to Wales.

Woodford's portfolio at Invesco included a variety of biotech firms at different stages of development and there were failures as well as successes. One of his companies, Phytopharm, collapsed in 2013 after trials showed that its treatment for Parkinson's disease showed no benefit to patients.[47] Another Invesco-backed firm, Proximagen, a spin-out from Kings College London which focused on diseases of the central nervous system, was floated on AIM in 2005 and sold in 2012 to Upsher Smith of the US for £223m.[48]

Woodford's approach stemmed from a personal conviction, not necessarily shared by all his Invesco colleagues, that British academic science ought to be translatable into economic gain, and that the only way of achieving that gain was through patient, long-term support for firms that had the potential to grow into profitable businesses. That his style of investing was more personal than corporate was made clear at the end of 2013 when he announced that he was leaving Invesco to set up his own firm.[49] Woodford said that he would continue to invest in biotech in the same way as he had done at Invesco. With his initial portfolio, published in June 2014, he allocated the bulk of the fund to large, well-established companies—some in healthcare, some in other

[46] In May 2015 the largest investor in ReNeuron, with 26 per cent of the shares, was Neil Woodford's recently established investment fund, Woodford Investment Management. Invesco and Abingworth each held about 5 per cent, and the Wales Life Sciences Investment Fund 11 per cent. A further fund-raising in July 2015 increased Woodford's share to 35.5 per cent.
[47] In 2013 Phytopharm was delisted from the main market and transferred to AIM as a cash shell; it was then acquired in a reverse merger by another Invesco-backed firm, IXICO (chaired by Andy Richards), which was a supplier of clinical trial and imaging services for research into dementia and other neuro-degenerative diseases.
[48] Proximagen was backed by the IP Group.
[49] At Invesco the management of Woodford's funds was taken over by his long-time colleague, Mark Barnett, and there was no indication that Barnett would adopt a radically different investment approach. Invesco remained the principal shareholder in Imperial Innovations and IP Group, alongside the new Woodford fund, and in several of the companies which Woodford had supported.

sectors—but slightly less than 10 per cent of the fund was invested in smaller biotech firms.[50]

If Woodford was to persuade others to back early-stage firms and support them over a long period, he needed to demonstrate that his approach was capable of producing some outstanding successes—that is, firms that listed on the stock exchange, preferably in London rather than New York, brought drugs to the market, and gave investors a handsome return. An important step in this direction came in March 2014, when Circassia, which had been backed by Invesco since 2008, was floated on the London Stock Exchange's main market.

The London Market Opens Up

The Circassia flotation was a remarkable event, not only because it ended the long IPO drought on the London Stock Exchange, but also because of its size. The £202m that it raised from investors made it the biggest IPO in the history of the UK biotech sector, far exceeding the £50m raised by Ark in 2004. The IPO gave Circassia a market value of £583m.

This was a company which did not expect to generate revenue until 2017 and had twenty employees at the time of the flotation. Its lead drug, a treatment for cat allergy, was in Phase III trials and it had three other unpartnered drugs which were in Phase II. It had a proprietary 'Tolero-Mune' platform, developed at Imperial College, which in principle could be applied to a wide range of allergies and to other immune disorders. It had a strong financial position and an experienced management team. The chief executive, Steve Harris, had been chief financial officer at Powder-Ject, one of the most successful of the first generation UK listed biotechs. His head of research, Dr Rod Hafner, had also been a senior manager in that firm.

Circassia was exceptional in that more than half the shares were owned by three investors, Imperial Innovations, Invesco, and Lansdowne Partners, which were committed to support the IPO.[51] Although the IPO was a success, reflecting some revival in investor appetite for drug discovery firms, the shares finished the year lower (by 10 per cent) than the IPO price, while in the rest of

---

[50] In 2015 he launched a new investment trust, Woodford Patient Capital Trust. When fully invested, this trust would have some 50–100 holdings, of which 75 per cent would be early-stage and early-growth firms, in biotech and other sectors, and the others would be mature, dividend-paying stocks.

[51] Woodford later invested in Circassia through his own fund, Woodford Asset Management, which in June 2015 held a 14 per cent stake in the company.

Table 5.2. Selected biotech and biotech-related IPOs in London in 2014

Company	Market	Amount raised (£m)	Market cap at IPO (£m)
Circassia	Main	202.0	583.0
Horizon Discovery	AIM	40.0	120.5
Abzena	AIM	20.0	77.5
4D Pharma	AIM	16.6	36.6
C4X Discovery	AIM	11.0	31.0
Venture Life*	AIM	4.2	26.4
Midatech Pharma	AIM	32.0	74.2

* Venture Life is a manufacturer of food supplements, anti-ageing cosmetics and medical devices
Source: Based on figures supplied by Bio Trinity, Peel Hunt, March 2015.

the world the stock price trend post-IPO was generally upward.[52] No other biotech firm followed Circassia on to the main market in 2014.

Also on the menu for London investors on the day of the Circassia IPO was a follow-on placing by a more established, revenue generating company, Vectura, which was already listed on the main market. Vectura issued new shares to the value of £52m and the offer was three times over-subscribed; Invesco had a 13 per cent shareholding in Vectura but the rest of the shares were widely held.[53]

Circassia aside, the other IPOs held in London in 2014 were on AIM, with several biotech or life science-related firms joining the junior market (see Table 5.2). Some of them, such as Horizon Discovery and Abzena, were technology providers, not drug discovery firms, and part of their appeal to investors was that they were already generating revenue, or were about to do so, and did not depend on high-risk, early stage research.[54] This was a business model which many UK-based investors preferred, even if the potential returns might be lower than in a pure drug discovery firm. However, there were also some successful IPOs by firms which depended wholly or mainly on drug discovery.

In this category the most spectacular debut was that of 4D Pharma, which floated in February 2014; by March 2015 its shares had risen by no less than 247 per cent, thanks in part to good progress with its two lead drugs, Blautix,

---

[52] Chris Morrison and Riku Lähteenmäki 'Public Biotech in 2014—the numbers', *Nature Biotechnology* 33, 7 (July 2015) pp. 703–9.

[53] At the same time Vectura announced the purchase of a German company, Activaero, whose inhalation products complemented those of Vectura, for €130m.

[54] The principal early backer of Horizon Discovery was Jonathan Milner, a former academic scientist—he had worked in the Cambridge laboratory led by Professor Tony Kouzarides—who had created Abcam and built it into a highly profitable supplier of antibodies and reagents. Abcam had listed on AIM in 2005 and had become one of its most highly valued companies. Milner was a prominent member of the Cambridge business angel community.

Table 5.3. Selected follow-on capital issues in 2014

Company	Market	Amount raised (£m)
Summit	AIM	21.0
Proteome Sciences	AIM	5.0
Synairgen	AIM	44.1
Vectura	Main	52.0
SkyePharma	Main	112.1
Silence Therapeutics	AIM	11.4
Oxford BioMedica	Main	21.6
Retroscreen Virology	AIM	33.6

Source: Based on figures supplied by Bio Trinity, Peel Hunt March 2015.

a treatment for irritable bowel syndrome, and Thetarix, a treatment for paediatric Crohn's disease. One of the founders and senior executives was David Norwood, who had started the IP group and had founded or co-founded several biotech firms.[55]

Meanwhile, some of the biotech firms which were already listed were able to raise follow-on capital during 2014 (Table 5.3). They included a few old-timers which had been in existence for 15 years or more and had managed to stay afloat through the ups and downs of the stock market, the vicissitudes of their drug development programmes and several failed partnerships.

One of the survivors was Oxford BioMedica, which raised £22m in 2014. This was a gene therapy firm which since its move from AIM to the main market in 2001 had been a disappointment for investors. The two founders, Alan and Sue Kingsman, had built the business on a novel approach to lentiviral vectors, a tool for delivering genetic material into cells, but it took longer than expected to convert the technology into marketable products.

Oxford BioMedica also had a cancer vaccine, Trovax, licensed from Cancer Research Technology and partnered with Sanofi, but that arrangement came to an end in 2009 after negative results from clinical trials. A new partnership was then negotiated with Sanofi, covering the use of Oxford BioMedica's gene delivery technology for the treatment of ocular diseases.

The chief executive, John Dawson, appointed in 2008, added a new source of revenue when he acquired a factory in Oxford (previously owned by British Biotech) to make lentiviral vectors for sale to other companies. This led to an agreement with Novartis, whereby the Swiss group gained access to Oxford BioMedica's technology for use in developing a novel leukaemia drug;

---

[55] In April 2015 three of the largest investors in 4D Pharma were Woodford Investment Management, Invesco, and Lansdowne Partners.

Novartis made an initial payment of $14m, which included taking a 2.8 per cent stake in the British firm.[56] The shares rose sharply after the Novartis deal, raising hopes that the patience of investors would at last be rewarded.[57]

By 2014 Oxford BioMedica had been in existence for nearly 20 years. Another survivor from the same generation was SkyePharma, a drug delivery firm; it had its headquarters in the UK but its main R&D base in Switzerland. It had listed on the main market in 1996 and grown through multiple acquisitions. This firm had taken the unusual step of raising funds through convertible bond issues; most biotechs rely on equity for their financing. When several of its programmes suffered delays—including the development of its most important drug, flutiorm, a treatment for asthma—the firm was in a financial bind, and the conversion of bonds massively diluted the equity investors. After a change of management in 2006 financial controls were tightened and the company renegotiated the terms of the bond debt. However, repeated delays in bringing flutiform to the market and high levels of debt continued to constrain the business.

In 2012 flutiform was finally approved by European regulators and by 2014 sales of this drug had reached €72m; with the drug now approved in thirty-one countries analysts were predicting sales of €350m by 2020. SkyePharma was also benefiting from growing income from other products.[58] In 2014 the rising share price enabled the company to raise new capital of £112m, which was used to repay its long-standing bond debt.[59] The company had finally shaken off the legacy of the past.

As a revenue-generating firm with drugs on the market SkyePharma was in a different category from high-risk drug discovery firms, and hence potentially attractive to generalist investors. But, as Peter Grant, chief executive, commented, 'It has certainly helped that there has been a change in UK stock market sentiment towards the biotech sector as a whole. A large capital-raising exercise like the one we did would have been much more difficult a few years earlier when all small drug development firms were out of favour.'[60]

[56] *Financial Times*, 11 October 2014.

[57] A supportive shareholder since the early 2000s had been the M & G Recovery Fund, which held just over 17 per cent in 2015; the other big shareholder, with 28 per cent, was a specialist healthcare investor, Vulpes Investment Management, which had been an investor since 2005.

[58] These included Exparel, an injectable drug for post-surgical analgesia, marketed by SkyePharma's former injectable business, which it had sold in 2007.

[59] In March 2015 SkyePharma was awarded the 'Best Performing Share' and 'Turnaround of the Year' awards at the 2014 PLC awards; these awards are given to main market-listed companies outside the FTSE 100.

[60] Interview with Peter Grant, 26 June 2015.

The Unquoted Firms

While public market investors were showing a greater interest in UK biotech, there was also an encouraging flow of funds into unquoted firms, with much of the money coming from non-British investors (Table 5.4). As noted earlier, the biggest funding in 2014, for Adaptimmune, was backed by US venture capitalists. There was also an American interest in NuCana BioMed, a Scottish-based firm that was developing anti-cancer medicines based on a novel technology, known as ProTide, invented by Professor Chris McGuigan at Cardiff University. NuCana's £34m Series B round was led by Sofinnova Ventures from the US.[61] The two founders, Chris Wood and Hugh Griffiths, had links to the US through their earlier company, Bioenvision, which they had sold to Genzyme in 2007 for $345m.[62]

In the case of Adaptimmune, the fund raising was quickly followed by an IPO on NASDAQ. Whether any of the others would do the same, either in New York or London, depended on what view their owners took about the long-term potential of the business and on how soon they wanted to cash out. The alternative was a trade sale, and there was no lack of buyers eager to snap up promising UK biotech firms (Table 5.5).

One such firm was Heptares. Spun out from the MRC's Laboratory of Molecular Biotechnology in 2007, Heptares had innovative science and experienced management, and it was backed by highly regarded and mostly

Table 5.4. Principal venture capital funding rounds in 2014

Company	Amount raised (£m)	Funding round	Principal investors
Bicycle	20.00	Series B	Atlas Venture, Novartis Venture Fund, SR One, SV Life Sciences, Astellas Venture Management
Kymab	23.80	Series B	Syncona, Bill and Melinda Gates Foundation
NuCana	33.98	Series B	Sofinnova Ventures, Sofinnova Partners, Alida Capital, Morningside Ventures, Scottish Investment Bank
Cell Medica	50.00	Series B	Imperial Innovations, Invesco, Woodford Investment Management
Adaptimmune	67.60	Series A	New Enterprise Associates, OrbiMed Advisors, Wellington Management, Fidelity Biosciences
NightstaRx	12.00	Seed	Syncona

Source: Bio Trinity, Peel Hunt.

[61] Sofinnova Ventures had historic links with Sofinnova Partners in Paris but was run independently. The Paris-based Sofinnova was the largest shareholder in Nucana and it also supported the Series B round.

[62] Bioenvision, based in Edinburgh but listed in New York, had developed a treatment for paediatric acute leukaemia, clofarabine, which was partnered with Ilex in the US. Ilex was later acquired by Genzyme, and the partnership with Bioenvision continued.

Table 5.5. Pharma/biotech acquisitions, 2010–2015

Acquirer	Target	Upfront deal value ($m)	Overall deal value ($m)	Date
Johnson & Johnson (US)	Respivert	104	120	2010
Kyowa Hakkin Kirin (Japan)	ProStrakan	475	475	2011
Jazz (US)	EUSA	650	700	2012
Upsher Smith (US)	Proximagen	347	553	2012
AstraZeneca (UK)	Spirogen	200	440	2013
Ipsen (France)	Syntaxin	37	170	2013
Biogen Idec (US)	Convergence	200	675	2015
Sosei (Japan)	Heptares	180	400	2015

Source: HBM Pharma-Biotech M & A Report.

non-British investors. Its focus was on treatments for Alzheimer's disease and schizophrenia. Heptares was seed funded by MVM, and MVM also participated in the Series A round in 2009, alongside Clarus Ventures from the US and the Novartis Option Fund. The Series B round in 2013 was co-led by the Stanley Family Foundation, a leading US funder of research into neurological diseases, supported by Clarus Ventures and Takeda Ventures.

At this stage the lead drug was about to enter Phase II clinical trials, but more funding would be needed before Heptares was ready for an IPO. Various funding routes were considered, including a Series C round and the sale of one of the company's promising drug candidates. The alternative was to sell the company, and this was the preferred option, given the dilution that would follow from another private round and the uncertain prospects for any future IPO; an early sale also suited the interests of the most time-constrained investors.

In 2015 Heptares was sold to Sosei of Japan for an initial payment of $180m, with further milestone payments up to a maximum of $220m.

While Sosei clearly saw great value in Heptares and was determined to make the best use of the scientific assets it was acquiring, the outcome was disappointing for those who had seen Heptares, as one observer put it, as 'Britain's answer to Vertex'.

The comparison between Circassia (an IPO in London), Adaptimmune (an IPO on NASDAQ), and Heptares (a trade sale) illustrates the three exit options which were open to British firms and their owners in 2014 and 2015. How the balance between them would evolve over the next few years depended partly on US investor sentiment towards the biotech sector (which was beginning to weaken towards the end of 2015), partly on whether other biotechs would attract sufficient support from UK-based institutional investors to permit a listing in London, and partly on the success of the new generation of pioneering firms.

The UK Biotech Sector in 2015

In mid-2015 UK biotechnology was in a healthier state than it had been for many years. As Steve Bates, chief executive of the BioIndustry Association, remarked, the UK had emerged from a 'nuclear winter' in life science financing.[63] Yet some of the problems that had dogged the sector since the turn of the century had not gone away. Despite the revival in biotech IPOs—mainly on AIM rather than the main market—the financing gap between British and American firms working in the same therapeutic area was still very wide. An extreme case was that of Silence Therapeutics, one of several firms that raised fresh capital in 2014. Silence was a specialist in RNA interference; its lead drug, a treatment for pancreatic cancer, was about to enter Phase II trials. The leading American firm in RNAi was Alnylam, which since its creation in 2003 had raised capital, through partnerships with Big Pharma and from public market investors, on a scale that dwarfed Silence's fundraising efforts. In July 2015 Alnylam was valued in the stock market at $10bn, Silence at just over £200m.

The Circassia IPO had been a bold move on the part of Invesco and the other shareholders. Yet Circassia when it came to the market was still a loss-making firm some years away from generating revenue.[64] Its success or failure would have a big impact on investor attitudes to the sector as a whole. Circassia was also a test of the investment philosophy which Neil Woodford had been pursuing for several years. He believed that if investors were prepared to support a promising biotech firm for long enough, and to nurture it in the private sector until it was ready to go public, there was a fair chance that it would do well.

In terms of market capitalization, thanks to Circassia and GW's success on NASDAQ, the structure of the UK biotech sector was less unbalanced than it had been before the financial crisis (Table 5.6). But for policy makers the absence of mid-sized drug discovery firms, filling the gap between Big Pharma and small biotech, was still a matter of concern, and that concern was heightened by the uncertain future of Britain's two big pharmaceutical companies, GlaxoSmithKline and AstraZeneca.

GSK was in the throes of an extensive restructuring aimed at concentrating on fewer therapeutic areas; the biggest move was an asset swap with Novartis,

[63] *Financial Times*, 14 October 2014.

[64] In May 2015 Circassia bought Aerocrine, a profitable Swedish supplier of diagnostic equipment for asthma and other airway diseases, and Prosonix, a British firm developing generic drugs for asthma and COPD; the acquisitions were funded through a £275m share placing. These deals reduced Circassia's dependence on novel drug development and were seen by some analysts as moving it closer to the cash-generating speciality pharmaceutical model. Steve Harris, chief executive, spoke of Circassia's ambition 'to be the next Shire'. *Financial Times*, 12 May 2015.

Table 5.6. Market capitalization of leading British drug discovery companies, July 2015

Company	Market capitalization (£m)
GlaxoSmithKline	65,840.0
AstraZeneca	53,870.0
Shire Pharmaceuticals*	32,602.2
BTG	2,564.8
GW Pharmaceuticals	1,711.3
Circassia	856.1
Vectura	731.3
Adaptimmune	704.0
4D Pharma	590.5
Vernalis	303.8
SkyePharma	291.4
Oxford BioMedica	236.6
Silence Therapeutics	202.9

* Shire Pharmaceuticals was founded in the UK but is now head-quartered in Ireland.
Source: Rx Securities.

whereby it acquired the Swiss company's vaccine and consumer products businesses in return for what was regarded as its sub-scale cancer portfolio.[65] At AstraZeneca Pascal Soriot, who had been recruited from Roche in 2012 to be chief executive, was trying to make up for several years of under-performance by strengthening its pipeline and forging alliances with biotech firms. In 2014 AstraZeneca came close to being taken over by Pfizer, a deal which some government ministers saw as a threat to the future of the British pharmaceutical industry.[66] Although Soriot fought off the bid (thanks in part to support from Invesco, one of the principal shareholders), the company was facing patent expiries on two of its best-selling drugs. Its future depended on how quickly it could bring new products to the market and how successful those products would be. In mid-2015 AstraZeneca had fourteen drugs in late-stage development, including some promising cancer drugs.

In the remarks quoted at the start of this chapter, Sir John Bell, the Oxford academic who was one of the government's life sciences advisers, drew attention to the dearth in the UK of large, independent biotech firms comparable to Celgene and Gilead in the US.[67] Past failures, he said, had scarred UK investor sentiment towards the high-risk sector; more success stories were needed to draw money from the big institutions. Sir John himself, who had been

[65] In 2015 Sir Andrew Witty, GSK chief executive, announced a change of strategy, which involved a shift away from high-risk, high-reward prescription drugs and a greater emphasis on vaccines and consumer healthcare. *Financial Times*, 12 May 2015.

[66] Also in 2014 AbbVie, the US pharmaceutical company, which had been split off from Abbott Laboratories in 2012, announced a takeover bid for Shire, but this did not arouse political opposition because Shire had few connections to the UK. The bid was later withdrawn.

[67] *Financial Times*, 30 March 2015.

involved in the creation of Avidex in 1999, joined the board of Immunocore in 2015; this was a company, he said, which had a chance to become the kind of mid-sized drug maker the UK lacked.

A sign of changing investor attitudes was the successful flotation of several technology transfer companies, following the example set by Imperial Innovations and the IP Group.[68] These included two US-based firms, Allied Minds and PureTech, which listed on the London Stock Exchange in 2014 and 2015, evidently in the belief that this was a business model which UK investors understood, and that there were plenty of opportunities for exploiting discoveries made in British universities.

There were also new initiatives from the universities. In 2015 Oxford joined with a group of City institutions to set up a £300m fund—Oxford Sciences Innovation—to develop science and technology businesses. The investors included some of the leading supporters of UK biotech—the Wellcome Trust, Lansdowne Partners, Woodford Investment Management, and the IP Group. It was seen as an attempt to replicate on a broader and larger scale what the IP Group had done 15 years earlier when it formed a partnership with the university's Department of Chemistry; David Norwood, former head of the IP Group, was appointed chairman of the new Oxford fund.

All this was good for start-up and early-stage firms. What remained uncertain was how many of them would stay independent for long enough to achieve an IPO, and if they did so whether they would float in London or New York. The shift to NASDAQ by GW, Adaptimmune, and others was not necessarily bad for the UK economy, since in most cases the research function on which the business depended remained in the UK; British investors are still able to hold shares in these firms and purchase new ones. But there was a danger that the centre of gravity in these firms would gradually shift to the US, leaving the research operation in the UK somewhat isolated and insecure. There was a perception among policy makers that the full value of public investment in biomedical research would not be realized unless it was linked to companies which, even if a large part of their business was in the US, had their headquarters and much of their top management, as well as their research laboratories, in the UK.

That 2014 and 2015 marked a significant improvement in the sector's fortunes was not in doubt. But as a 10-year review commissioned by the BioIndustry Association pointed out, past experience had shown that success could quickly turn to disaster.[69] If the recovery was to be sustained, then either

---

[68] Both Imperial Innovations and IP Group raised new capital in 2014 and 2015, £150m in the first case and £128 in the second. Advent Life Sciences, led by Raj Parekh, raised £146m in a new fund in 2014.

[69] *UK Biotech: A 10 year horizon*, published by BioIndustry Association and Evaluate Ltd, June 2015.

more sizeable, cash generating companies would need to come to the market or some beacons of hope for investors—firms with successful new medicines—would need to emerge. This in turn would encourage the return of specialist investors who had played such a central role in the growth of biotech in the US but were largely absent in the UK. While the prospects were much brighter than in the years before the financial crisis, there were still gaps in the UK ecosystem which had to be filled.

6

Learning from the US

An International Comparison

We want Germany to be a leading location for biotechnology in Europe. In fact we want it to be the best. But don't get me wrong. I'm not talking about coming out on top in Europe here. What I'm talking about is a Europe competing with the United States for the markets of the future. I want Europe as a whole to hold its own in the face of global technological competition. And I am convinced that this old continent of ours has what it takes to succeed. Dr Jürgen Rüttgers, German Minister of Education, Research and Technology, May 1998[1]

When the US Office of Technology Assessment (OTA), a government agency set up by Congress to analyse science and technology issues, examined the world biotech scene in 1984, it identified five foreign countries as the major potential competitors of the US in this field.[2] These were: Japan, Germany, the UK, Switzerland, and France. In four of these countries—Switzerland being a partial exception—biotech had been targeted by governments as a sector that merited special encouragement and support. Part of the purpose of the OTA study was to examine steps that the US might take to strengthen its competitive position.

The study was written at a time of growing concern in the US about the competitiveness of American industry. Japanese companies had made inroads into the US market for cars and TV sets, and there were fears that the same thing could happen in biotech. The OTA writers warned that US leadership could be eroded unless the decline in government funding for basic life science research was halted. They also called for more support for generic

---

[1] Speech given at the 1998 European Life Sciences Conference in Amsterdam, quoted in *BioCentury* 6, 39 (4 May 1998).
[2] *Commercial Biotechnology: An International Analysis* (Washington, DC: US Congress, Office of Technology Assessment 1984) pp. 7–8.

applied research, especially in bioprocess engineering and applied microbiology, and for clearer rules on health, safety, and environmental protection.

In the light of subsequent events, it is clear that the OTA report underestimated both the strength of the US in biotech and the difficulties which other countries would face as they tried to promote their biotech sectors. As discussed in the last three chapters, the UK made a promising start, but failed to build on it. This chapter examines the other four countries named by the OTA, focusing on the obstacles that held back the growth of their biotech firms and assessing how far those obstacles have been removed.

Japan

In discussing Japan, the OTA study noted that there was a consensus within the Japanese government about the importance of biotechnology to the future of the economy. A number of collaborative projects between government and industry had been launched, with the motto 'catch up, get ahead'. This targeting policy, together with the manufacturing skills of established companies such as Takeda, Suntory, and Ajinomoto, was likely to make Japan a formidable competitor.[3]

This judgement proved to be wide of the mark. It was true that the rise of Genentech and the other US pioneers had alerted the Japanese authorities to the potential significance of biotechnology. In December 1980, a few weeks after the granting of a patent for the Boyer/Cohen invention of recombinant DNA techniques, senior representatives from Japanese companies attended a hurriedly called meeting of the Committee on Life Sciences of the Japan Federation of Economic Organisations (the Keidanren). The aim was to discuss how best to ensure Japanese participation in the new industry and to resist the threat of US domination. The meeting agreed that counter-measures were necessary, and this view was shared by officials in the Ministry of International Trade and Industry (MITI).[4]

In 1982 MITI set up an Office of Biotechnology Promotion. Following its well-established policy for supporting new industries, MITI organized cooperative research programmes involving pharmaceutical and chemical companies. However, at a time of difficulty for the Japanese economy and a growing budget deficit, MITI was unable to secure political backing for a large, government-funded biotechnology programme. In 1984 the government spent no more

---

[3] *Commercial Biotechnology: An International Analysis,* p. 505.
[4] Gary R Saxonhouse, 'Industrial policy and factor markets: biotechnology in Japan and the United States', in *Japan's High-technology Industries: Lessons and Limitations of Industrial Policy,* edited by Hugh Patrick and Larry Meissner (Seattle: University of Washington Press 1986) p. 99.

than $35m on biotechnology, less than in Germany, the UK, or France, and far less than in the US.[5]

Even if the government had spent more, there was a fundamental difference between Japan and the US which slowed down the exploitation of the new techniques. Biotech in the US was driven by new, entrepreneurial firms, mostly founded or co-founded by academic scientists. Japan, by contrast, had traditionally relied on large, established companies as the principal source of early-stage innovation.[6] There was no tradition of academic scientists leaving universities to start commercial ventures. Equally, it was rare for scientists in industry to break away from their companies to start new firms; most of them expected to stay with their first employer for their entire career.

Thus the development of biotechnology depended on existing companies, and the obvious candidates, as far as medical applications were concerned, were the pharmaceutical companies. However, although Japan had a large pharmaceutical industry, the leading companies were less internationally oriented than their European and American counterparts and less committed to research. This was partly due to protectionism; the industry was insulated from foreign competition in the domestic market. Another factor was a reimbursement system, run by the Ministry of Health and Welfare, which provided little incentive for the development of innovative drugs.[7]

In these circumstances the pharmaceutical companies were unlikely to make a big investment in a novel technology where the risks were high and the rewards uncertain. Much of the activity in the early years came from companies in other industries—principally brewing, food manufacturing, and chemicals—which saw biotechnology as sufficiently close to their existing skills to provide a basis for diversification.[8] For example, Kirin, the largest Japanese brewer, made an agreement with Amgen in 1984 whereby it obtained the right to market Epogen in Japan. Kirin and other non-pharmaceutical companies had some success, but they had difficulty recruiting the necessary scientific skills, and they focused much of their efforts, not on innovative projects, but on well-established technologies which offered the prospect of an early return.[9]

---

[5] Saxonhouse, 'Industrial policy', pp. 106–7.

[6] Robert Kneller, *Bridging Islands, Venture Companies and the Future of Japanese and American Industry* (Oxford: OUP, 2007) p. 1.

[7] Robert Kneller, 'Autarkic drug discovery in Japanese pharmaceutical companies: insights into national differences in industrial innovation', *Research Policy*, 32 (2003) pp. 1805–27. L. G. Thomas, *The Japanese Pharmaceutical Industry, the New Drug Lag and the Failure of Industrial Policy* (Cheltenham: Elgar, 2001), pp. 65–7.

[8] Yumiko Okamoto, 'Paradox of Japanese biotechnology: can the regional cluster development approach be a solution?' Department of Policy Studies, Doshisha University, Unpublished paper, December 2008.

[9] Kneller, *Bridging Islands*, pp. 192–205.

By the 1990s it was clear that, far from catching up with the US in biotechnology, Japan was falling further behind. The Japanese model for creating new industries, based on collaboration between established companies under the guidance of MITI, was not working. Policy makers now looked to the American model as a means of recovering lost ground.[10]

An early priority was to make it easier for scientists to start new businesses. In 1998 the government introduced a law to promote the transfer of technology from universities into industry; by the end of the decade most leading universities had set up technology transfer offices, partly subsidized by the state. Other laws gave universities and public research institutes the right to own and exploit patents on their discoveries, and allowed university staff to hold jobs in a private company while retaining their university positions.[11]

There were also improvements in access to finance. In 1991 the Tokyo Stock Exchange launched an over-the-counter market for smaller firms, although a minimum level of profitability was still required—a disadvantage for R&D focused biotech firms which had no products on the market. In 1995, under pressure from MITI, a second OTC market was set up which allowed high-technology firms without a profit record to go public, and this was taken further in 1999 with the launch of what became known as the Mothers Exchange (Market of the High Growth and Emerging Stocks), partly modelled on the London Stock Exchange's AIM.

One of the first to take advantage of these changes was Ryuichi Morishita, an assistant professor at Osaka University who had spent several years as a post-doctoral fellow at Stanford, working on hepatocyte growth factor (HGF), a gene involved in organ growth and regeneration through its role in stimulating the development of new blood vessels. The gene's potential to grow back damaged vascular systems in diabetic patients had commercial potential, and when Morishita returned to Japan he tried without success to interest Japanese pharmaceutical companies in his idea.[12] In 1999 he founded his own company, MedGene, later renamed AnGes MG, raising just enough money to keep the business going until he negotiated an alliance with Daiichi Pharmaceuticals, which bought the marketing rights to his technology. AnGes MG went public on the Mothers Market in 2002, and by 2007 about seventeen biotech start-ups had followed its example, although only three of them were drug discovery firms.[13]

---

[10] Leonard H. Lynn and Reiko Kishida, 'Changing paradigms for Japanese technology policy: SMEs, universities and biotechnology', *Asian Business and Management* 3 (2004), pp 459–78; Maki Umemura, 'Crisis and change in the system of innovation: the Japanese pharmaceutical industry during the Lost Decades, 1990–2010', *Business History* 56, 5 (2014) pp. 816–44.

[11] Michael J. Lynskey, 'Bioentrepreneurship in Japan: institutional transformation and the growth of bioventures', *Journal of Commercial Biotechnology*, 11/1 (3 October, 2004) pp. 9–37.

[12] David Pilling, 'AnGes moves out of the dog house', *Financial Times*, 28 September 2003.

[13] Okamoto, 'Paradox of Japanese biotechnology'.

Meanwhile changes had been taking place in the Japanese pharmaceutical industry. Under pressure from international pharmaceutical companies and from foreign governments, restrictions on the ability of non-Japanese firms to market their drugs in Japan were removed, and controls on inward investment were liberalized. Foreign companies, instead of licensing their drugs to Japanese firms, could now set up their own sales organizations in Japan. The cross-licensing model on which most Japanese companies had previously relied—licensing their innovative drugs to foreign companies in return for the right to sell overseas-developed drugs in Japan—was no longer viable. They needed to build up their business outside Japan and to compete more effectively in biologics.

What followed was a series of acquisitions by Japanese companies in the US and Europe (Table 6.1). Most of the targets were biotech firms, and the acquisitions had a dual purpose: to get control of big-selling or potentially big-selling biologics and to acquire sales organizations through which they could market their own drugs abroad. By this time the leading Japanese companies, thanks in part to changes in the patent laws and to a greater reluctance on the part of the regulators to approve non-innovative drugs, had increased their spending on research and some of them were achieving impressive results. In cholesterol, for example, some of the statins marketed by Western companies, such as AstraZeneca's Crestor (rosuvastatin), were based on technology licensed from Japan.[14]

Yet the major Japanese companies remained focused on traditional pharmaceutical R&D approaches and were still lagging behind in biologics. A rare success was a novel antibody therapy for Castleman's disease and rheumatoid

Table 6.1. Major Pharma/biotech acquisitions by Japanese companies, 2007–2015

Acquirer	Target	Price ($m)	Year
Eisai	MGI Pharma (US)	3,900	2007
Takeda	Millennium (US)	8,200	2008
Shionogi	Sciele (US)	1,400	2008
Dainippon Sumitomo	Sepracor (US)	2,600	2009
Astellas	OSI Pharma (US)	4,000	2010
Takeda	Nycomed (Norway)	13,680	2011
Daiichi Sankyo	Plexxikon (US)	805	2011
Takeda	URL Pharma (US)	800	2012
Otsuka	Astex (US)	886	2013

Source: HBM Partners, Pharma/Biotech M & A Report.

[14] Other successes were Takeda's Actos (pioglitazone), a competitor to Glaxo's Avandia in diabetes drugs, Eisai's Aricept (donepezil), a treatment for Alzheimer's disease, and Otsuka's Abilify (aripiprazole), an antipsychotic drug which was partnered with Bristol-Myers Squibb in the US.

arthritis, tocilizumab (Actemra), which stemmed from collaboration between Chugai and Osaka University.[15] But this was an exception. According to Robert Kneller, a close observer of the Japanese pharma/biotech industry, very few biochemically innovative and medically significant drugs originating in Japanese universities have been developed by large Japanese pharmaceutical companies.[16] Instead, this role has been performed either by foreign companies or by smaller Japanese firms.

For example, oxaliplatin (Eloxatin), a widely used chemotherapy drug, was discovered in Nagoya City University; Debiopharm, a Swiss firm, undertook development after no Japanese pharmaceutical partner could be found. A lung cancer drug, crizotinib (Xalcori), discovered by scientists in Jichi University, was quickly picked up by Pfizer. Among small Japanese pharmaceutical firms, Yoshitomi Pharmaceutical Industries worked with Professor Tetsuro Fujita in Kyoto University to develop the small molecule drug fingolimod (Gilenya), first used as an immunosuppressant for kidney transplant patients and later relaunched as a treatment for multiple sclerosis; it was licensed to Novartis in 1994.

The most spectacular Japanese success was that of Professor Tasuku Honjo, also based in Kyoto, who discovered a new way in which tumours suppress the immune system. After many years of collaboration with Ono Pharmaceutical a novel antibody drug that blocked this inhibition mechanism, nivolumab (Opdivo), was developed and approved by Japanese regulators in 2013 and then by the FDA in 2014. Bristol-Myers Squibb of the US was Ono's partner for worldwide marketing of this drug, which was expected to achieve blockbuster status.[17]

While Ono's achievement was remarkable, it did not alter what Kneller sees as a continuing weakness in the Japanese innovation ecosystem, its over-dependence on large, established pharmaceutical companies which have a poor record in producing breakthrough therapies. In the US many of these breakthroughs have come from small biotech firms. Firms of this sort have been slow to develop in Japan, partly for the reasons discussed earlier, but partly also because the big Japanese pharmaceutical companies have until recently had a low regard for local biotech firms; they were not seen as attractive partners.[18]

---

[15] Roche bought a controlling interest in Chugai in 2001 and acquired a licence to Actemra in 2003.

[16] This is documented by Robert Kneller in his study of the origins of all the new therapeutic drugs approved by the FDA over a recent 10-year period. Robert Kneller, 'Importance of new companies for drug discovery: origins of a decade of new drugs', *Nature Reviews Drug Discovery* 9 (November 2010) pp. 867–82.

[17] In 2005 Ono had partnered with Medarex in the US to develop this drug. Medarex was bought by Bristol-Myers Squibb in 2009.

[18] Kneller, 'Autarkic drug discovery in Japanese pharmaceutical companies'.

That attitude has begun to change in recent years as some Japanese biotechs have forged alliances with European and American pharmaceutical companies. For example, PeptiDream, a specialist in peptide therapeutics co-founded by Professor Hiroaki Suga at the University of Tokyo, has partnerships with several Western companies, one of which, Novartis, has a minority stake in the Japanese firm. PeptiDream listed on the Mothers Market in 2013.

There has also been growing interest on the part of local and international investors in Japanese biotech, helped by the evident determination of the Abe government to stimulate the growth of high-technology firms. In 2012 the government launched the START programme which used government funding, together with management advice from the private sector, to help academics start new businesses.[19] Biotech ventures were among the first START-supported projects.

Investors have been impressed by the progress made by Japanese academic scientists in developing new treatments for intractable illnesses. In 2012 Professor Shinya Yamanaka of Kyoto University was awarded (jointly with John Gurdon from the UK) the Nobel Prize for physiology or medicine, based on his research into regenerative medicine using pluripotent stem cells. The announcement prompted a surge in the share prices of firms that were working in areas close to Professor Yamanaka's research, including AnGes MG.

By 2015, thanks to the reforms made over the previous 20 years, Japan had moved some way towards creating an independent biotech sector. Yet there was still some reluctance on the part of academic scientists to form new firms, and a shortage of business angels and venture capitalists who had the skill and experience to identify and coach the best teams. Japan's experience in biotech shows how difficult it is to change long-established institutions and management practices. A business model that had worked well in industries such as automobiles and electronics proved unsuitable for an industry which depended on multiple sources of experimentation, and on the willingness of scientists and entrepreneurs to embark on high-risk ventures with no guarantee of a return.

Germany

The most obvious difference between Germany and Japan when biotechnology came on the scene was the strength of the German pharmaceutical industry.[20] Companies such as Hoechst and Bayer ranked among the world

[19] OECD Science, Technology and Industry Outlook (Paris: OECD, 2014) pp. 360–2. START is an acronym for Programme for Creating Start-ups from Advanced Research and Technology.
[20] This was based on the early lead which firms such as Hoechst and Bayer had established in the late nineteenth century in applying synthetic organic chemistry to pharmaceuticals. German

leaders in the industry and they had a more distinguished record of innovation than their Japanese counterparts. But there were also some similarities between the two countries. They both had financial systems which were poorly equipped to support entrepreneurial firms in new industries. In both countries the norms of academic life, and the way career structures were organized in universities and companies, discouraged scientists from breaking away from their employers to create new businesses.

Banks were the principal source of external finance for German companies. Only the largest firms were listed on the stock market; most medium-sized companies chose to stay private. Although small firms could obtain funds from state-owned organizations at the federal or regional level—the largest was the Kreditanstalt fur Wiederaufbau (Kfw)—there was little support for new entrepreneurial firms with no track record. Between 1945 and 1980 IPOs played almost no role in Germany. Partly because of the absence of an exit through the stock market, the venture capital industry was far less developed than in the US.[21]

Investment in biotechnology was also held back by political opposition to genetic engineering research, especially from the Green Party. This was one of the reasons for Hoechst's decision, in 1981, to negotiate an agreement with the Massachusetts General Hospital in Boston; in return for a $70m contribution to the cost of a new molecular biology department, the German company would have exclusive access to all commercially exploitable discoveries coming out of Hoechst-sponsored research.[22] This event came as a shock to policy makers, highlighting the fact that Germany was lagging behind both in molecular biology research and in its commercial applications. The Hoechst deal prompted the government to establish Gene Centres in Cologne, Heidelberg, and Munich, although it was not until the Genetic Engineering Law of 1993 that the restrictions on genetic engineering research were fully removed.

Despite this unhelpful environment, some German scientists were watching with interest the progress of Genentech and its imitators in the US. One of the first to follow their example was Turkish-born Metin Colpan, founder of Qiagen, which was to become the most successful of the German biotech firms. While taking his PhD at the University of Darmstadt, Colpan became

scientists such as Robert Koch and Paul Ehrlich had also been pioneers in the use of 'old biotechnology' to produce biological drugs.

[21] Stefanie A. Franzke, Stefanie Grohs, and Christian Laux, 'Initial public offerings and venture capital in Germany', in *The German Financial System*, edited by Jan P. Krahmen and Reinhard H. Schmidt (Oxford: OUP, 2004).

[22] Thomas Wieland, 'Ramifications of the "Hoechst shock": perceptions and cultures of molecular biology in Germany', Working Paper, Munich Centre for the history of science and technology, August 2007.

convinced of the need for new methods of extracting the pure nucleic acid needed for genetics research. He developed a novel technique for isolating ultra-pure plasmid DNA from bacteria without the use of toxic chemicals or expensive equipment. After failing to get backing from pharmaceutical companies he obtained finance from an American venture capital firm, Alafi Capital, and from Techno Venture Management (TVM), one of the few German VCs willing to back start-up firms.[23] In 1984 Colpan established Qiagen as a Netherlands-registered company, with its main operations in Hilden, near Dusseldorf.

Qiagen earned revenue by supplying products and services to other firms, a less risky approach than drug discovery, and this model was adopted by several other German companies.[24] In 1992 two Qiagen managers, Karsten Henco and Ulrich Aldag, broke away to form their own company, Evotec, based on a high-throughput screening technology developed by Professor Manfred Eigen from the Max Planck Institute. (Eigen had won the Nobel prize for chemistry in 1967.) Initial finance for Evotec came from TVM and a group of private investors.

By this time a venture capital industry was taking shape, supported by several non-German firms. A London-based investor, Alex Korda, provided seed funding for Morphosys, founded in 1992 to exploit research on monoclonal antibodies at the Max Planck Institute for Biochemistry in Munich; other investors included Atlas Venture, then based in the Netherlands, and 3i from the UK. The company was co-founded by Simon Moroney, a New Zealander who had worked as an academic in the US and for an American biotech firm, and Andreas Plückthun from the Max Planck Institute. Another Munich firm, Medigene, was founded in 1994 by scientists from the Gene Centre at the University of Munich to develop gene therapy treatments for heart disease and cancer.

What was still missing was a receptive stock market for biotech IPOs. Qiagen, which had grown rapidly since its foundation, chose to list on NASDAQ in 1996.[25] The flotation was a success—US investors bought more than 60 per cent of the shares on offer—but the chairman, Carsten Claussen, said that the

---

[23] Like Advent in the UK and Sofinnova in France, TVM was established with the support of the Boston VC firm, TA Associates.

[24] Steven Casper suggests that the German preference for service-based or platform-based business models rather than drug discovery reflected the long-standing attachment in much of German industry to incremental rather than radical innovation. Steven Casper, 'Institutional adaptiveness, technology policy and the diffusion of new business models: the case of German biotechnology', *Organisation Studies*, 21, 5 (2000) pp. 887–914. Hannah E. Kettler and Steven Casper, 'The road to sustainability in the UK and German biotechnology industries', Office of Health Economics (London, July 2000).

[25] For tax and legal reasons Qiagen listed on NASDAQ through a Dutch holding company. Its head office has remained in Holland.

need to go to the US was 'a sad experience' since it reflected the extreme difficulty of raising equity capital in Germany.[26]

Qiagen's IPO came at a time when the Kohl government was seeking to inject new dynamism into the German economy by promoting high-technology industries. Biotechnology was one of the chosen sectors, and the aim was to overtake the UK and make Germany the European leader in this field (see the quotation at the head of this chapter).[27] In 1995 the Federal Ministry of Research launched the BioRegio Competition, through which regions could apply for Federal funding, matched by investment from the private sector, to provide start-up and early-stage biotech firms with subsidies covering up to 50 per cent of their research costs. Federal funds could also be used to establish incubator laboratories around universities and technology parks. The three winners of the competition were Munich, Heidelberg, and Cologne (a special award was also made to Jena in East Germany).

The effect of the scheme was to stimulate a surge of biotech start-ups and to encourage more venture capitalists to take an interest in the sector. In 1997 the Frankfurt Stock Exchange launched the Neuer Markt, which, like AIM in London, had less demanding rules than the main market and allowed loss-making firms to obtain a listing. At the start the Neuer Markt was a big success. Many start-up and early-stage firms, in biotech and other high-technology sectors, went public. Their shares soared as investors were caught up in dot-com mania and the general enthusiasm for the industries of the future. But the boom proved unsustainable. While share prices in most of the world's leading stock markets, including NASDAQ, fell back sharply in 2000 and 2001, the collapse of the Neuer Markt was more extreme, and exposed the frailty of many of the firms that had gone public. The reputation of the market was 'defaced by profit warnings, insolvency filings and adverse publicity ranging from misleading information to spectacular mismanagement and suspected cases of fraud and insider trading'.[28] The Neuer Markt was closed in 2003.

The years following the market crash saw a contraction in the number of German biotech firms and in the number of venture capitalists willing to invest in the sector.[29] As one investor remarked: 'Hopes for a new investment and shareholder culture in Germany have been dashed. Instead, we got greedy

---

[26] Andrew Fisher, 'Venturing across the pond', *Financial Times*, 23 July 1996.

[27] Karen E. Adelberger, 'A developmental German state? Explaining growth in German biotechnology and venture capital', BRIE Working Paper 134, 1999; Dirk Dohse and Tanja Staehler, 'BioRegio, BioProfile and the rise of the German biotech industry', Kiel Working Papers No 1456, October, 2008; Susan Giesecke, 'The contrasting roles of government in the development of biotechnology industry in the US and Germany', *Research Policy* 29 (2000), pp. 205–23.

[28] Bertrand Benoit, 'Rise and fall of a high-tech brand', *Financial Times*, 14 August 2001.

[29] Siegfried Bialojan and Julia Schüler, 'Commercial biotechnology in Germany: an overview', *Journal of Commercial Biotechnology* 18, 1 (September 2003) pp. 15–21.

to make a killing and many invested in the hope to make quick money. They have been disappointed and now the market is dead.'[30]

Not everyone was so pessimistic. Some argued that the combination of government subsidy and easier access to private finance had been a catalyst for German biotech and had helped to strengthen the two main clusters in Munich and Heidelberg. What mattered now was whether the sector could regain the confidence of investors by producing companies capable of profitable growth.

By 2003 the sector had recovered some stability. Some of the firms which had gone public in the late 1990s, such as Morphosys and Evotec, were able to raise new capital. Morphosys was emerging as the largest and strongest of Germany's biotech firms, using its HuCal antibody platform to generate licence revenue from partners and as the basis for its own drug development. The biggest partnership, worth up to $1bn, was signed with Novartis in 2007, giving Morphosys the financial strength to build its own proprietary pipeline.

There was one IPO on the Frankfurt Stock Exchange in 2004, three in 2005, and six in 2006. However, despite the size of the German economy, the amounts of money raised were small, and, as in the UK, some ambitious firms preferred to try their luck in the US. In 2006 Micromet, a cancer specialist which had been spun out from the University of Munich in 1993, merged with NASDAQ-listed CancerVax, a loss-making but cash-rich company. Having moved to the US, Micromet raised some $300m to develop its antibody technology platform, but its main laboratory remained in Munich.

Within Germany finance was available from government support schemes at the federal and regional level. In 2005 the government set up the High-Tech Grunderfonds, a public–private venture capital fund whose remit was to supply seed finance for start-up firms in advanced technology; the bulk of the funding came from the government, with smaller contributions from the private sector. Two years later the government launched the GO-Bio competition, through which academic scientists could apply for funds to convert their ideas into a business. This was followed by the Leading Edge Cluster Competition, through which selected regions were given funding over a 5-year period to strengthen network-based collaboration between firms and universities in science-based industries.

This period also saw the emergence of wealthy private investors with an interest in biotech. One was Dietmar Hopp, who had co-founded SAP, the software company, in 1972. He left the company in 2005, and put most of his wealth into a charitable foundation, which invested in a range of biotech firms. The other big private investors were the Strüngmann brothers, Andreas

---

[30] *BioCentury*, 21 October 2002.

and Thomas, who had created Hexal, one of the world's largest generic drug manufacturers. When they sold the company to Novartis in 2005, they put their wealth to work in biotech.

Most other investors, however, both private and institutional, had been badly burned by the boom-and-bust of 2000–01, and had lost confidence in German biotech. The world financial crisis of 2008–09 added to the sector's woes, and it was slow to recover. In its 2011 report on German biotech Ernst & Young reported that the traditional venture capital model was no longer working in Germany.[31] A growing number of German biotech firms withdrew from drug discovery and refocused on providing services or developing platforms that could generate revenue through licensing deals with pharmaceutical companies.

Only in a few cases were German firms able to follow the classic route, from seed funding through venture capital to IPO. Of the fifteen largest venture rounds in Europe in 2014, only one was for a German firm, compared to five from the UK and four from Switzerland. Two German firms had IPOs in 2014, but neither chose to list in Frankfurt. Affimed Therapeutics, which had spun out from the Cancer Research Centre in Heidelberg in 2000, was backed by several non-German investors (including OrbiMed in the US) before floating on NASDAQ in 2014. Probiodrug, which was developing a novel treatment for Alzheimer's disease (its backers included the venture capital arm of Biogen Idec), floated on Euronext Amsterdam in 2014.

The quality of German science continued to attract interest from foreign investors and from Big Pharma acquirers. In 2012 Micromet was bought by Amgen of the US for $1.1bn, the biggest trade sale in the history of German biotech. Micromet was by then a NASDAQ-listed company, but its research base was in Germany and Amgen made it clear that it would retain the Munich laboratory. The attraction of Micromet for the acquirer lay partly in its lead drug, a treatment for blood malignancies which was in Phase III clinical trials, and partly in the German company's innovative antibody platform.

There were hopes that the Micromet sale would revive the interest of German investors in local biotech firms but the financing situation remained difficult. Although new firms were still being created (there was no lack of seed capital, much of it from government sources), Germany's competitive position in biotechnology was far weaker than the one that it had traditionally enjoyed in pharmaceuticals. In 2013 it ranked sixth among European countries in the number of drugs that had reached Phase II or Phase III trials

---

[31] Ernst & Young German Biotechnology Report 2011 (Ernst & Young Gmbh Mannheim April 2011).

(Figure 6.1). Among listed firms, only Morphosys could be regarded as a genuine success (Table 6.2).

One possible response was to accept that chasing after the US in biotech was unlikely to be productive, and that Germany's main contribution should be in the provision of tools, services, and technologies, rather than drug discovery. The opposing view was that, because of the inadequate supply of venture capital and the absence of IPOs, Germany was failing to obtain full value

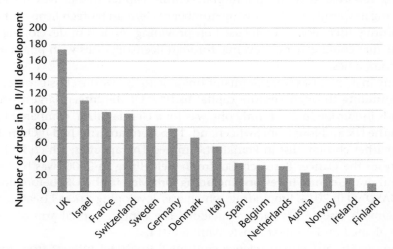

Figure 6.1. Number of drugs in Phase II and III clinical trials in 2013
Source: Ernst & Young, and Medtrack.

Table 6.2. Top fifteen European biotech firms measured by market capitalization, July 2015

Company	Country of origin	Market capitalization (£m)
Actelion	Switzerland	10,779.0
Genmab	Denmark	3,252.3
BTG	UK	2,564.8
SOBI	Sweden	2,225.5
GW Pharma	UK	1,711.3
Cosmo	Italy	1,673.0
Galapagos	Belgium	1,282.7
Morphosys	Germany	1,224.4
Circassia	UK	856.1
Bavarian Nordic	Denmark	836.0
Basilea	Switzerland	823.4
Cellectis	France	758.9
Vectura	UK	731.3
DBV	France	722.8
4D Pharma	UK	590.5

Note: If Shire were included in the table, it would rank far above Actelion, but in this as in most other analysts' reports Shire is generally grouped with Big Pharma companies such as GSK and AstraZeneca.
Source: Rx Securities.

from its investment in biomedical research. As Siegfried Bialojan, head of Ernst & Young's European Life Sciences Centre in Mannheim, remarked, 'Germany has huge innovation potential, thanks in part to subsidies, but because of the gaps in the food chain we have given away a lot of value. It makes no sense to invest so heavily in research if it is disconnected from value creation'.[32]

Given the mixed results of the subsidy programmes launched over the previous 20 years, no one was suggesting a large injection of public money into the sector. The focus now was on how to get private sector investors more fully engaged. Among the ideas which the industry was discussing with government officials was the introduction of tax incentives to encourage institutional investors to allocate a small part of their portfolio to early-stage firms in high technology industries.

If investors were to shift away from traditional industries such as cars and engineering, there might also be a need to rethink what was widely regarded as the over-favourable tax treatment of debt financing vis-à-vis equity. An increase in the supply of risk capital was seen as essential to the future growth of the biotech sector.[33]

France

France has a long and distinguished history of excellence in medical research, dating back to Louis Pasteur's work on the germ theory of disease in the mid-nineteenth century. The Institut Pasteur, set up as a private foundation in 1888, played a major role in molecular biology in the 1950s and 1960s; three of its scientists, François Jacob, André Lwoff, and Jacques Monod, won the Nobel Prize for Medicine or Physiology in 1965 for their research on the mechanisms of gene regulation in microorganisms.[34] Yet France was slow to commercialize the new discovery techniques that came to the fore in the 1970s and 1980s. As in Japan and Germany, there were institutional blockages which took a long time to remove.

First, scientific research was dominated by state-owned institutions, of which by far the largest was the Centre National de la Recherche Scientifique (CNRS). The CNRS was charged with conducting research that promoted the interests of the French nation, but the orientation of its staff was almost wholly academic, with little interest in transferring the results into industry.

---

[32] Interview with Siegfried Bialojan, 29 April 2014.

[33] Momentum nutzen, Politische Signale setzen für Eigenkapital und Innovation, Deutscher Biotechnologie-Report 2015, Ernst & Young.

[34] Michel Morange, *A History of Molecular Biology*, translated by Matthew Cobb (Cambridge, MA: Harvard University Press, 1998).

They had few links either with the universities, whose primary function was teaching, or with the *grandes écoles*, which trained the elite of French youth for careers in government or in large companies that were closely associated with the state. Scientists who worked in CNRS and the other state laboratories, including INSERM, which handled medical research, enjoyed job security and attractive salaries; there was no incentive for them to plunge into the high-risk world of commercial start-ups. Status and standing in the scientific hierarchy were 'negatively associated with engagement with the private sector'.[35]

Second, French industrial policy in the first three decades after the war was dominated by *grands projets*, through which large companies, some of them state-owned, were required by the government to carry out programmes in strategic industries such as defence and aerospace, and in infrastructure sectors such as telecommunications and nuclear power. The government sought to foster national champions which could hold their own against competition from the US; there was little interest in small firms as a source of innovation.

This was not a favourable environment for the creation of US-style biotech firms. One that attempted to break the mould was Transgene, founded in 1979 by two leading biologists, Pierre Chambon and Philippe Kourilsky, to exploit the work on gene transfer technology that they had developed at the Institut Pasteur. The prime mover was Robert Lattès from Paribas, who persuaded three large industrial companies to provide some of the initial funding.[36] He hoped that Transgene would demonstrate that high-risk ventures in advanced technology could succeed in France as well as in the US, but few others followed his example.

The foundation of Transgene came at a time when the US biotech sector was growing rapidly, prompting fears in government that France might be missing out in a potentially important industry. In 1980, under the presidency of Giscard d'Estaing, the government asked a senior civil servant, Jean-Claude Pelissolo, to make proposals for an action programme in biotechnology. The premise of the report was that biotechnology was still at an early stage and it was not too late for France to catch up.[37] The report urged the government to improve the transfer of technology between academia and industry and to initiate a series of pilot programmes, funded by the state, which would bring together teams from the public and private sectors.

[35] Michelle Gittelman, 'National institutions, public–private knowledge flows and innovation performance: a comparative study of the biotechnology industry in the US and France', *Research Policy* 35 (2006) pp. 1052–68.

[36] *Financial Times*, 25 May 1984.

[37] Margaret Sharp, 'Biotechnology in Britain and France', in *Strategies for New Technology: Case Studies from Britain and France*, edited by Margaret Sharp and Peter Holmes (London: Philip Allan, 1989) pp. 139–40.

Some of these proposals were taken up by François Mitterrand's Socialist government which entered office in 1981. Biotechnology was one of several sectors targeted by the government as part of an ambitious industrial policy, aimed at strengthening France's position in high-growth industries. Yet the initiatives launched over the next few years had little impact. The BioAvenir programme, launched in 1992, involved collaboration between the principal research organizations, including INSERM and the Institut Pasteur, and Rhône-Poulenc, the country's largest pharmaceutical company.[38] Although Rhône-Poulenc was able to initiate research projects in areas of its choice, the programme did little to strengthen the biotech sector.[39] According to one critic, 'There was no start-up generated in France with BioAvenir. It was rather one of these big projects *à la française*, which only benefited big pharma'.[40]

A few scientist-entrepreneurs started biotech firms during this period, backed by a growing venture capital industry. But, as Ernst & Young remarked in its 1996 report on European biotech, most of the French companies were little more than contract research houses, lacking the funds to undertake long-term research.[41] 'Biotechnology in France', according to the report, 'has not developed anywhere near to its full potential'. The authors attributed the dearth of entrepreneurial bioscience firms partly to lack of finance, partly to French academic culture—the focus of most academics on building their scientific reputations and their isolation from industry. An academic observer, Michelle Gittelman, noted how few scientists left public laboratories to work in the private sector. Only about forty to fifty were leaving public sector employment every year, out of a total of 26,000, and most of them were taking temporary leave. 'A primary reason why there is no biotechnology industry in France', she wrote, 'is that the people needed to form one—academic scientists—have not made themselves available to the task.'[42]

What was needed was a reorientation of industrial policy in favour of entrepreneurs, and this began to happen during the 1990s.[43] In 1997 the government introduced Fonds Communs de Placement dans l'Innovation

---

[38] Anne Branciard, 'France's search for institutional schemes to promote innovation: the case of genomics', paper presented at the workshop on innovation, industry and institutions in France, Paris, CEBREMAP, February 2000.

[39] In 1999 Rhône-Poulenc merged its pharmaceutical interests with those of Hoechst to form Aventis. In 2004 Aventis was acquired by Sanofi, which became the largest French pharmaceutical company.

[40] Sabine Louet, 'French genomics setup questioned', *Nature Biotechnology* 18 (2000) pp. 375–6.

[41] Ernst & Young, *European biotech 96, volatility and value*, p. 56.

[42] Michelle Gittelman, 'Mapping national knowledge networks: scientists, firms, and institutions in biotechnology in the United States and France', PhD Dissertation, University of Pennsylvania (2000) p. 92.

[43] Philippe Mustar and Philippe Laredo, 'Innovation and research policy in France (1980–2000) or the disappearance of the Colbertist state', *Research Policy* 31 (2002) pp. 55–72. For a review of the recent evolution of French innovation policy, see OECD Reviews of Innovation Policy: France, December 2014.

(FCPI), which gave tax incentives to investors holding shares in start-up firms. Two years later the government adopted the 'Law on innovation and research to promote the creation of innovative technology companies'. This allowed scientists to participate in commercial ventures without losing their academic posts, and encouraged universities and public research laboratories to set up incubators to support the creation of new firms.[44] The government also set up a FF100m BioAmorçage fund to provide seed capital for biotech firms; it was a way of making up for the lack of US-style business angels. As Antoine Papiernik of Sofinnova, the leading French venture capital firm, remarked at the time, 'France is still at the first generation of biotechnology entrepreneurs, in contrast with the US which is reaching its third generation.'[45]

In 1996 the Paris Bourse opened a new market, the Nouveau Marché, for early-stage firms. Partly modelled on AIM in London, it imposed less rigorous listing rules than the main market. Two of the first biotech firms to list on the new market were Transgene and Genset, a genomics firm that had been founded in 1989.[46] Genset listed on NASDAQ at the same time.

Like the Neuer Markt in Germany, the Nouveau Marché was hit hard by the 2000–01 boom-and-bust, but it had not attracted as many speculative issues as the Neuer Markt, and the stronger companies were able to switch to the main market. In 2000 the Paris Bourse joined forces with the stock exchanges in Brussels and Amsterdam in a cross-border exchange called Euronext, although this did not involve full integration; it was not a step towards a Europe-wide stock market as envisaged in the short-lived EASDAQ scheme. In 2005 a junior market, Alternext, was introduced, and it attracted several smaller biotech firms.[47]

Meanwhile the government continued to improve the tax environment for entrepreneurs. The 2003 Innovation Plan established a new legal status for business angels and provided a range of incentives for entrepreneurial firms. Firms that were categorized as *jeunes entreprises innovantes* were given tax exemptions. To qualify for this status firms had to be less than 8 years old and have fewer than 250 employees and a turnover of less than €50m; their spending on research had to amount to at least 15 per cent of sales.

---

[44] Philippe Mustar and Mike Wright, 'Convergence or path dependence in policies to foster the creation of university spin-off firms? A comparison of France and the United Kingdom', *Journal of Technology Transfer* 35, 1 (February 2010) pp. 42–65.

[45] Quoted in Sabine Louët, 'New law to boost French biotech industry', *Nature Biotechnology*, 17 (1999) p. 1055.

[46] Both these companies disappointed investors. Genset switched from genomics to drug discovery and was acquired by Serono in 2002. Transgene passed into the control of the Mérieux family but remained a public company.

[47] In 2007 Euronext merged with the New York Stock Exchange group and was renamed NYSE Euronext.

The next step was an ambitious plan for creating industrial clusters—*pôles de compétivité*—designed to encourage collaboration at the local or regional level between companies, research institutes, and educational institutions. Although the programme was criticized at the start for being too large and too unfocused, over time it served as a useful vehicle for promoting dialogue between academic scientists and business. Several of the biotech clusters, notably those based in Lyon and Marseilles, helped to spawn new firms.

While the emphasis on small firms marked a change from the national champion era of the 1960s and 1970s, it did not imply a retreat from state intervention in industry. The election of Nicolas Sarkozy as President in 2007 led to renewed activism in the field of industrial policy.[48] In addition to intervening to save companies in difficulty, President Sarkozy launched a strategic investment fund—Fonds Stratégique d'Investissement (FSI)—to invest in companies which were deemed to be critical to the competitiveness of the French economy. Biotechnology was one of the sectors in which the FSI took a special interest, and it supported several firms during the world financial crisis when funding from private sources was scarce.

In 2012 FSI's responsibilities were widened through a merger with OSEO-Anvar, the government innovation agency, to create Banque Publique d'Investissement (Bpifrance). This brought under one roof a range of instruments that could be used to support firms at different stages of development. These included a biotech-focused seed capital fund, Innobio, which was managed by Bpifrance but had as co-investors several pharmaceutical companies, including Sanofi, GSK, and Roche.

One of the companies that benefited from FSI support was Innate Pharma, a specialist in immunotherapy for cancer and inflammatory diseases. Innate was founded in 1999 in Marseilles by a group of immunologists led by Hervé Brailly.[49] An early success, validating Innate's technology, came in 2003 when it signed a research partnership with Novo Nordisk of Denmark; the Danish company also took a shareholding in Innate. The prospects were good enough to allow the company to float its shares on Euronext Paris in 2006. However, its financial position remained fragile and when the banking crisis struck in 2008 its share price fell sharply, making it difficult to raise new capital. Innate might have been forced either to cut back its operations or to look for a buyer, but in 2009 it secured an injection of capital from FSI. FSI became the second

---

[48] Jonah D. Levy, 'The return of the state? The economic policy of Nicolas Sarkozy', paper presented at the annual conference of the American Political Science Association, Washington DC, 2–5 September 2010.

[49] Brailly had earlier worked for a Marseilles-based biotech firm, Immunotech, whose main business was diagnostics. It was bought in 1995 by Coulter of the US, which was later acquired by Beckman.

largest shareholder, with 13 per cent of the shares, just behind Novo Nordisk with 15 per cent.

Innate was working in a therapeutic area which was attracting strong support from investors and from Big Pharma. In 2011 Innate formed a partnership with Bristol-Myers Squibb for the development and commercialization of one of its lead drugs; the agreement involved an initial payment of $35m and milestones potentially amounting to $430m. The share price rose sharply after the announcement, and although Innate was listed in Paris rather than New York, US investors started to take an interest in the company. In 2013 OrbiMed and other US healthcare investors subscribed to a €20.3m private placement, increasing the proportion of Innate's share capital held by US institutional investors to more than 25 per cent.

A few months later Innate was able to acquire a highly differentiated compound from Novo Nordisk and to raise €50m to fund its development. In 2015 Innate went on to strike a licensing deal for that compound with AstraZeneca; the deal was worth $250m upfront and a further $1bn in milestones. By then the market capitalization of Innate was over €700m and US institutional investors held more than 35 per cent of the company's share capital.

Innate was not the only French firm to tap into the US investor community. In 2014 and 2015 DBV Technologies, a specialist in anti-allergy drugs, and Cellectis, which was operating, like Innate, in the 'hot' area of immuno-oncology, launched successful IPOs on NASDAQ, while retaining their listings in Paris.[50] (Both these firms had Bpifrance as one of their major investors.) Genfit, which was developing a treatment for a liver disease known as non-alcoholic steatohepatitis or NASH (another 'hot' therapeutic area), also raised capital from US investors, and was a likely candidate for NASDAQ.

These and other firms were benefiting from what had become a far more favourable financing environment. Three key elements were the FCPI scheme, the R&D tax credits (which had been introduced in the 1980s but were simplified and made more generous in 2008) and the special provisions under the young innovative enterprise scheme. FSI, now Bpifrance, had provided a valuable additional source of funding, supplementing what was available from venture capital firms such as Sofinnova. There was also growing interest in the sector from local investors, private as well as institutional. Euronext Paris could not hope to rival NASDAQ as a source of capital, but the Paris market was attracting an increasing number of life science IPOs

---

[50] DBV raised $93m when it floated on NASDAQ in October 2014. In July 2015 it raised a further $245m in one of the largest public offerings made by a European company in the US. *Financial Times*, 15 July 2015.

Table 6.3. Life sciences IPOs on Euronext Paris, 2005–2014

Year	No of IPOs	Amount raised €m
2005	2	37
2006	3	49
2007	5	140
2008	1	—
2009	0	—
2010	7	10
2011	6	92
2012	8	146
2013	7	153
2014	9	237

Source: France Biotech Faits Marquants 2015, May 2015. These figures include medtech, diagnostics and greentech as well as biotech IPOs.

(Table 6.3). In the first three-quarters of 2015 biotech or biotech-related firms raised substantially more money in IPOs and follow-on capital issues on Euronext Paris than in London.[51]

Another positive development was a more constructive attitude towards local biotech firms on the part of Sanofi, the dominant French pharmaceutical group.[52] Sanofi had been slow to shift away from the chemistry-based approach to drug discovery, and when it began to establish a position in biologics its main focus was on the US. (Sanofi's biggest move in the US was the purchase of Genzyme in 2011.) As the quality of science in French biotechs became more evident, Sanofi was more willing to consider local alliances. In 2015 Sanofi entered into a collaboration agreement with Innate for the development of new antibody drug conjugates, a novel approach to the treatment of cancer.

There are still weaknesses in the French innovation ecosystem, shared to some extent with other European countries: the absence of specialist healthcare investors, the dearth of evergreen venture capital funds that can take a view longer than 5–8 years, and continuing bureaucratic obstacles in transferring technology from academia into industry. France has yet to produce a really big success story, and there is the ever-present risk that failures in clinical trials might undermine investors' new-found confidence in the sector. But French biotech has clearly recovered from its sluggish start and looks set to continue to grow.

---

[51] Analysis of European biotech companies on the stock markets: US versus Europe, Biocom AG (November 2015).

[52] Sanofi had been founded in 1973 as part of Elf Aquitaine, the oil company. In 1999 it acquired Synthélabo to become France's second largest pharmaceutical company after Rhône-Poulenc. In the same year Rhône-Poulenc merged its pharmaceutical interests with those of Hoechst to form Aventis. In 2004 Sanofi acquired Aventis.

Switzerland

Of the smaller European countries, Switzerland seemed the best equipped to take the lead in biotechnology. It had a strong pharmaceutical industry, several universities and research institutes with expertise in molecular biology, and an array of banks, financial institutions, and private investors capable of raising capital for new companies. Yet Swiss business was slow to respond to the opportunities created by the Boyer/Cohen and Milstein/Köhler discoveries.

Although Biogen, one of the first and ultimately most successful biotech firms, was founded in Geneva in 1978 (it later moved to Boston), it did not spark off a wave of start-ups, as Genentech had done in the US. Charles Weissmann, a leading Swiss molecular biologist who was a member of Biogen's founding team, said later: 'In the late 1970s I tried to interest Swiss pharmaceutical companies in the upcoming genetic engineering technology and there was zero interest.'[53]

During the 1980s and early 1990s the three big pharmaceutical companies—Hoffmann La Roche, Ciba-Geigy, and Sandoz[54]—began to take biotechnology more seriously, although much of their activity was in the US, either through setting up research laboratories or by forming links with US firms, as Roche did with Genentech and Ciba-Geigy with Chiron. The leading Swiss player in biotech in those years was Serono, an old-established pharmaceutical firm controlled by the Bertarelli family which had moved from Italy to Switzerland. Its biggest success was a treatment for multiple sclerosis, Rebif (interferon beta 1a), launched in 1997; it was a competitor to another interferon-based drug, Avonex, from Biogen in the US.

As for new entrants, there was not a strong tradition in Switzerland of academic scientists starting their own businesses;[55] breakaways from Big Pharma were the most likely source of new firms. The most successful of the newcomers was Actelion, formed in 1997 by a group of ex-Roche scientists led by Jean-Paul Clozel and his wife Martine. While working in Roche's cardiovascular department they had discovered that the endothelium, a thin layer of cells lining the inside of blood vessels, could be a contributory factor in heart failure and other diseases, and this was the area on which the new firm was focused.

Actelion was set up without financial support from Roche, but in 1998, after Roche had decided to withdraw from research on endothelium, it agreed to

---

[53] R. James Breiding, *Swiss Made, the Untold Story behind Switzerland's Success* (London: Profile Books, 2012) p. 241.

[54] Ciba and Geigy had merged in 1971. In 1996 Ciba-Geigy and Sandoz merged to form Novartis.

[55] Charles Weissmann's role in the founding of Biogen is a notable exception.

license two compounds on which it had been working, tezosentan and bosentan, to Actelion. The first venture funding came from Sofinnova and Atlas Venture (other firms, including 3i from the UK, came in later), and Actelion was listed on the Swiss Stock Exchange in 2000.

One of the two compounds, bosentan, was developed as a treatment for pulmonary arterial hypertension. Given the trade name Tracleer, it was approved in the US in 2001 and in Europe in 2002, and quickly increased its share of the market. (The other compound, tezosentan, targeted at acute heart failure, was not successful.) Actelion had other drugs in its portfolio, including Zavesca, a treatment for Gaucher's disease licensed from Oxford GlycoSciences in the UK, but Tracleer was the basis for Actelion's subsequent rise to become the most highly valued biotech firm in Europe. The speed of its ascent and the scale of its success made Actelion the bellwether of the European biotech sector; no other European firm has matched its achievement (Table 6.2).

Several other firms were spun out of the Swiss pharma companies over the next few years. Speedel, from Novartis, was developing small molecule-based blood pressure drugs; it was later re-acquired by Novartis for $880m after promising clinical results were published. Basilea, which acquired an anti-infective and dermatological portfolio from Roche, floated on the Swiss stock exchange in March 2004; it was the largest European biotech IPO in that year, raising almost twice as much as Ark, which had its London IPO in the same month.

Because firms with a Big Pharma background were pursuing drug programmes on which development work had already taken place, they were seen by investors as less risky than start-up firms founded by academic scientists. One investor suggested that firms coming out of Big Pharma 'helped to create a very solid, profitable base on which to build the younger, more high-risk companies'.[56]

During the 1990s the leading universities—principally Lausanne, Zurich, and Basel—established technology transfer offices, and several academic start-ups attracted support from Swiss and non-Swiss venture capital firms. For example, ESBATech, founded in 1998 to exploit an antibody fragment technology developed in the Institute of Molecular Biology at the University of Zurich, obtained funding from Clarus Ventures in the US and SV Life Sciences in the UK.

Closer links between academia and industry were encouraged by the government through SPP BioTech (Swiss Priority Programme Biotechnology), which was launched in the early 1990s; it led to the creation of Unitectra, a

---

[56] Comment by Nicholas Draeger, Adamant Biomedical Investments, in Swiss Biotech Report 2008 (www.swissbiotechreport.ch).

joint technology transfer organization for the universities of Basel, Berne, and Zurich. One of the firms supported by this programme was Cytos Biotechnology, which came out of the Federal Institute of Technology in Zurich, in 1995; its lead drug was a novel treatment for asthma.

By the early 2000s the Swiss biotech sector was gaining momentum. The venture capital industry was growing. HBM Partners, founded in 2001 by Henri B. Meier, who had previously been chief financial officer of Roche, was one of several Swiss venture capital firms that focused on biotech. The venture capital arms of Roche and Novartis were also active. At the end of 2003 Serono was the most highly valued European biotech firm, ahead of second-place Shire, and Actelion was in fifth position, just ahead of Celltech.

There were ten IPOs on the Swiss stock exchange between 2000 and 2007, including three from Italy—Newron, BioXell, and Cosmo Pharmaceuticals— attracted by the strong interest in biotech on the part of Swiss investors. Foreign interest in Swiss biotech was reflected in several partnership deals and acquisitions. In 2003 Biogen Idec of the US began collaborating with Fumapharm on the development of an oral small molecule drug, BG-12, for treating the relapsing/remitting form of multiple sclerosis. Three years later Biogen Idec bought the Swiss firm for an initial payment of $230m, rising to $500m if milestones were reached. The MS drug, dimethyl fumarate (Tecfidera), was launched in 2013 and became one of the US company's most successful products.

A merger which was to have less happy consequences was the 2007 takeover of Serono by Merck KGaA of Germany (no relation to Merck & Co. of the US).[57] Serono, despite diversification into other therapeutic areas, was still heavily dependent on Rebif, and with no big-selling drugs to follow it the Bertarelli family decided to put the company up for sale. Merck KGaA, for its part, was a medium-sized pharmaceutical company which needed to strengthen its position in biologics.

Serono aside, the Swiss biotech sector appeared to be riding high in the years leading up to the financial crisis. It was second only to the UK in the number of biotech-related drugs in clinical trials and in the market value of its listed companies. This had been achieved without direct government intervention of the sort that had taken place in the UK in 1980 (with the creation of Celltech) and in Germany in the mid-1990s. Yet the sector was more fragile than it seemed.

---

[57] Merck KGaA is one of the oldest German pharmaceutical companies. In 1891 it established a US subsidiary, which was confiscated by the US government during the First World War and later reconstituted as an independent US company, Merck and Co. Inc. There is no longer any connection between the German and US companies.

Of the Swiss biotechs that had listed on the Swiss stock exchange between 2000 and 2010 only Actelion and to a lesser extent Basilea could be regarded as genuine successes. Three others ran into problems. In 2009 Arpida, a Roche spin-out which had floated in 2005, suffered a setback when its lead drug, an antibiotic targeted at skin and soft tissue infections, was rejected by the FDA; it was forced to halt its research activities.[58] Two years later Addex laid off a quarter of its staff after its lead drug failed in clinical trials, although it was still hoping to take forward its treatments for Parkinson's disease and schizophrenia. In 2014 Cytos Biotechnology closed down all its research and development operations after the failure of its lead asthma therapy in Phase II clinical trials. Another blow came in 2012 when Merck announced the closure of the ex-Serono site near Geneva, with the loss of some 500 jobs; biologics research would be concentrated at Merck's main site in Darmstadt, Germany.

Even the star of the sector, Actelion, found itself under attack from activist investors who complained of the company's failure to produce follow-on drugs to offset the anticipated decline of Tracleer; in 2010 Tracleer accounted for 85 per cent of the company's revenues. At the end of that year the shares rose sharply in expectation of a takeover bid, and Actelion seemed likely to lose its independence. However, the company, still run by Jean-Paul Clozel, fended off the attack and by 2012, thanks to positive trial results for its follow-on drug to Tracleer, it had regained the confidence of shareholders.

Actelion had been an exceptional case; its lead drug was approved within 4 years of the company's foundation and quickly became a big seller. But if the sector was to retain the confidence of investors more firms needed to generate returns, if not on the Actelion scale, large enough to offset the well-known risks of putting money into biotech. Few such successes were forthcoming in the years following the financial crisis, with the result that the flow of IPOs on the Swiss stock exchange came to a virtual halt. It was not until the end of 2014, after a 5-year lull, that the IPO window opened wide enough to allow Molecular Partners to go public; this firm, a spin-out from the University of Zurich that specialized in small molecule drugs in oncology and ophthalmology, raised $100m, one of the biggest European IPOs in that year.

There were hopes that the Molecular Partners' IPO would be a catalyst for the rest of the sector, but the financing environment remained difficult. As one executive remarked, 'The classic model for financing biotechs...has become very challenging and is nowadays almost impossible. The strategic dilemma for today's drug developers is that investors are no longer willing to wait for ever or accept high attrition rates.'[59] The number of specialist health-care investors declined, and several banks closed down their biotech research

---

[58] Arpida was later acquired in a reverse merger by Evolva, a privately held Swiss firm.

[59] Harry Eelten, CFO, Cytos, in Swiss Biotech Report 2013.

departments. Venture capitalists such as HBM were increasingly focusing on firms with late-stage assets, as well as investing in firms that were already public; an increasing proportion of HBM's activities was in North America.

Thus Switzerland found itself in a situation not dissimilar to that of Germany. The quality of science in Swiss universities, and in some of the early-stage firms, was not in question. Indeed, an encouraging feature of the Swiss scene was the continuing willingness of non-Swiss pharmaceutical companies to look for partners or acquisition targets in Switzerland. GSK, for example, bought two Swiss vaccine firms, Okairos in 2013 for $324m and GlycoVaxyn 2 years later for $190m.[60] Several US-based venture capital firms, notably Versant Ventures, were active in the Swiss market.[61] The anxiety was that this activity by non-Swiss firms, together with the lack of interest on the part of Swiss investors, would encourage more Swiss biotechs to go to NAS-DAQ, and more Swiss scientists to move to the US. Auris Medical, which was developing treatments for tinnitus and other hearing disorders, went to NAS-DAQ in 2014, and others were expected to follow its example.

One hopeful development was the willingness of some wealthy individuals to invest in biotechnology. In 2011 two of the country's richest men, Ernesto Bertarelli (former owner of Serono) and Hansjorg Wyss (who had sold Synthes, a medical device manufacturer, to Johnson & Johnson in 2011 for $20bn) bought the ex-Serono site near Geneva. Their plan, which involved cooperation with the University of Geneva and the Swiss Federal Institute of Technology in Lausanne, was to turn the site into what they called Campus Biotech, a home for a new generation of biotech firms.

Like Dietmar Hopp and the Strüngmann brothers in Germany, Bertarelli and Wyss were much-needed supporters of Swiss biotech. But they could not make up for the reluctance of the big institutional investors, especially the pension funds, to invest in the sector. Given the traditional Swiss aversion to government intervention in industry, there was no question of injecting public funds into biotech, or of tax incentives to stimulate more investment. Any improvement in the funding environment would have to be driven primarily by the private sector.

In 2014 Henri Meier, one of Switzerland's leading venture capitalists, put forward the idea of what he called Future Fund Switzerland, through which pension funds would invest collectively in promising high-technology businesses. He pointed out that Swiss pension funds allocated only 6 per cent of their assets to alternative investments, and within that 6 per cent only 0.2 per

---

[60] Other deals included the sale of Neurimmune to Biogen Idec in 2010 for up to $395m and the sale of Covagen to Johnson & Johnson in 2014 for an undisclosed price.

[61] In 2014 Versant invested in three early-stage Swiss firms, Anokion, CRISPR Therapeutics, and PIQUR Therapeutics.

cent went to venture capital. His plan envisaged the creation by the pension funds of a new 'fund of funds' which would allocate money to specialized venture firms. Whereas no individual pension fund was large enough, or sufficiently risk-tolerant, to devote substantial resources to venture capital, the proposed Future Fund would provide a vehicle through which they could invest in a well-diversified and balanced portfolio; the fund would be a corner-stone, long-term investor in the Swiss venture capital industry. Under Meier's scheme, the government would not contribute to the Future Fund, but it would give its blessing to the proposal and encourage pension funds to participate.

Weaknesses in National Innovation Ecosystems

The events described in this chapter, together with the earlier chapters on the UK, underline the importance of four elements that are necessary for the establishment of a thriving biotech sector: well-funded universities with strong bioscience departments; effective arrangements for transferring academic discoveries into industry; an adequate supply of capital to support biotech firms at all stages of their development, from start-up through to commercialization; and an established pharmaceutical industry that provides experience, through the transfer of people, and resources, through partnerships and other commercial deals.

The second and third of these elements have been the principal focus of the institutional reforms and policy changes that have been implemented over the last few years. These reforms are national in scope and cannot in themselves offset some of the advantages enjoyed by US biotech firms: the massive support for biomedical research from the Federal government; easy access to a large and relatively price-insensitive domestic market which has been highly receptive to innovative drugs; and a large and sophisticated capital market which is well equipped to support young, high-technology firms.

In principle the European Union, given its size and its commitment to a barrier-free internal market, should be able to match these advantages, and some progress has been made in this direction.

On the research side, the European Commission has taken an interest in biotechnology since the early 1980s, when it introduced funding programmes for basic research in molecular biology, focusing on post-doctoral training and exchange.[62] Under the most recent Framework Programme (FP7) the

---

[62] Jacqueline Senker 'Biotechnology: the external environment', in *Biotechnology and Competitive Advantage: Europe's Firms and the US Challenge*, edited by Jacqueline Senker, co-ordinated by Roland van Vliet (Cheltenham: Edward Elgar, 1998) pp. 6–18.

Commission distributed €50bn for research (across all fields of science and technology) over the period 2007–13. FP7's successor, Horizon 2020, is projected to distribute €80bn across Europe in 2014–20.[63]

The life sciences, and in particular drug development, have been a major beneficiary of EU spending. For example, one activity within these programmes that spans both periods is the Innovative Medicines Initiative (IMI), which brings together over €2.5bn of European Union funding with an approximately equal contribution from industry. (The main industry contribution comes from Big Pharma rather than biotech firms, although smaller firms are able to join). The sums distributed overall are large; UK researchers were awarded around €7bn in the 2007–13 period, equivalent to the addition of a medium-sized research council to the UK funding landscape.[64]

As for reducing barriers to intra-European trade and investment, there has been some movement towards harmonization in regulatory arrangements, covering, for example, the use and transportation of genetically modified organisms that are essential for biotechnology research. The establishment of Europe-wide legislation on intellectual property rights, as well as on regulatory issues, has required coordination and agreement amongst all member states of the European Union. This has often been cumbersome and slow and in some cases the outcomes of legislation have been heavily criticized.[65] But EU-level cooperation has resulted in a useful pooling of technical expertise in European institutions.

The European Patent Office (founded in 1977) has become the preferred route for the examination of biotechnology-related patent applications (although national patent offices have not been replaced entirely). The European Medicines Agency (founded in 1995) has replaced national regulatory agencies as the authority for reviewing market authorization of new biologics. The European biotech sector has also benefited from the introduction of specific legislation that these institutions implement—the so called 'Biotech directive' which addresses the patentability of biotechnology related inventions (passed by the European Parliament in 1998) and orphan drug

---

[63] http://www.universitiesuk.ac.uk/highereducation/Documents/2013/BriefingHorizon2020 Budget.pdf.

[64] The amounts distributed to particular areas are difficult to ascertain due to international differences in definitions and data collection modalities (or lack thereof). See: National Audit Office, *Research and Development funding for science and technology in the UK*. Memorandum for the House of Commons Science and Technology Committee (London June 2013). http://www. nao.org.uk/wp-content/uploads/2013/07/Research-and-development-funding-for-science-and-technology-in-the-UK1.pdf.

[65] E. Richard Gold and Alain Gallocaht, 'The European Biotech Directive: Past as Prologue', *European Law Journal* 7, 3 (September 2001) pp. 331–66.

legislation (passed in 2000) with similar measures to the US Orphan Drugs Act of 1983.

Yet despite these changes the European Union is still a long way from establishing a genuinely integrated pharmaceutical market. National governments operate their own pricing and reimbursement arrangements, which tend to slow down the launch of new drugs. This is one of the reasons why European pharma and biotech firms generally look to the US as their primary market for launch; once approved by the FDA, their drugs can be brought to a market much larger than any national market in Europe and with a single launch.

The slow progress towards full integration is also reflected in the fragmentation of Europe's stock markets. After the failure of EASDAQ in the late 1990s there has been no move to create a Europe-wide stock exchange comparable to NASDAQ in the US. Supported as it is by a large number of specialist investors committed to biotech, NASDAQ is a far more attractive venue for biotech IPOs than any of the national European exchanges. There are no less than fifteen European exchanges on which biotech firms can list, although some of them are very small.

It is true that some European stock exchanges, especially Paris and London, have been able to attract US investors to their local biotech sectors; several of the leading European biotech firms, listed only on their national stock markets, now have a substantial proportion of their equity owned by US investors. What has been missing in some of the larger countries has been consistent support from local investors. Without that support ambitious firms looking for new capital tend to gravitate towards the US.

In the UK, as earlier chapters have shown, the reluctance of British institutional investors to support local biotech firms has long been a source of concern. A much discussed question is whether this phenomenon relates specifically to biotech, reflecting the exceptional risks associated with the sector, or is symptomatic of a broader weakness in the British financial system.

The British fund management industry, through which a large part of the nation's savings is channelled into equities, has been widely criticized for its bias towards short-termism—putting pressure on firms to maximize profits in the short term, at the expense of the long-term development of the business. This is seen as one of the reasons why young entrepreneurial firms in high-technology industries find it difficult to grow into large or even medium-sized companies.

The next chapter considers what light the biotech story sheds on this question.

7

The Financing of Biotech

If you want billion-dollar companies in the UK you have to have public markets ready to invest in them, and that is the biggest difference between the US and the UK. Where we might have had winners we have sold too early. Nigel Pitchford, chief investment officer, Imperial Innovations.[1]

'Where are Britain's Amazons, Apples or Googles?' This was a question posed by John Kay in his report to the British government, published in 2012, on UK equity markets and long-term decision making.[2] The purpose of the report was to examine whether there was a short-term bias in the British financial system which hampered the growth of high-performing businesses. The conclusion was that short-termism was indeed a serious problem, and that it stemmed in large part from the structure and organization of the fund management industry; fund managers were judged over too short a time horizon and had little incentive to focus on the long-term prospects of the companies in which they held shares. This in turn encouraged companies to look for ways of raising their share price in the short term, to the detriment of the long-term growth of the business.

Kay used GEC and ICI as examples of once-successful British companies which had been undermined by decisions aimed at maximizing shareholder value in the short term. In referring to Amazon, Apple, and Google he was making the point that few new, high-growth British firms had emerged in recent years to take the place of companies like GEC and ICI. He noted that one of the UK's most successful software firms, Autonomy, had been bought in 2011 by Hewlett-Packard of the US, with unhappy results for both companies.

---

[1] Interview with Nigel Pitchford, 20 June 2013. Before joining Imperial Innovations Pitchford had been a senior executive at 3i.
[2] John Kay, *The Kay Review of UK equity markets and long-term decision-making*, Department of Trade and Industry, Final Report (July 2012) para 1.27.

Kay might have used biotech to make the same point, asking why there were no British counterparts to Amgen, Biogen, or Gilead. In biotech as well as in information technology the UK's inability to create 'big gorillas', capable of competing with the world leaders, has long been seen as a weakness in the country's industrial structure, and the blame has often been put on the financial system.[3]

Of the UK biotech firms that were started in the 1980s and 1990s, only Shire (British by origin although no longer domiciled in the UK) lifted itself into the top rank of the global industry by size and profitability, and this company is in some respects a special case. Shire had been set up as a sales and marketing firm, not closely linked to the UK science base; it has grown through a string of acquisitions, and the bulk of its R&D is outside the UK. Among British firms that were founded on the basis of innovative science coming out of British universities, there were several which had ambitions to emulate the American leaders, but none of them achieved the valuations of their US counterparts or their commercial success.

Some of these firms, such as Celltech and Cambridge Antibody Technology, did build up sufficient value to be sold to larger firms at a high price. These takeovers, involving as they did the transfer of technologies to acquirers which were well equipped to exploit them (as well as giving shareholders a return on their investment), cannot be regarded as failures. Nevertheless, it has been argued that more UK biotech firms would have remained independent for longer, and made better use of their scientific assets (often acquired from publicly funded research), if the City had provided larger and more consistent support.[4]

Most of the world's big biotechs—entrepreneurial firms that were founded after the scientific breakthroughs of the 1970s and have grown into large, profitable businesses—are based in the US. Yet the American financial system is often criticized for much the same deficiencies as those that were highlighted by John Kay in his report on the UK; it has been compared unfavourably with the long-termist approach of countries such as Germany and Japan.[5] Larry Fink, chief executive of BlackRock, a prominent institutional investor, recently attacked the short-term demands of US capital markets, which, he

---

[3] Geoffrey Owen, *Where are the Big Gorillas? High technology entrepreneurship in the UK and the role of public policy* (Diebold Institute for Public Policy Studies, December 2004).

[4] George Cox, *Over-coming short-termism in British business: the key to sustained economic growth*, An independent review commissioned by the Labour Party (February 2013); Alan Hughes, *Short-termism, impatient capital and finance for manufacturing innovation in the UK*, Future of manufacturing project: Evidence Paper 16 (London, Government Office for Science, December 2013).

[5] Michael Porter, 'Capital disadvantage: America's failing capital investment system', *Harvard Business Review* (September/October 1993). This article was written at a time when admiration for Japanese long-termism was running strongly; this judgement had to be qualified in the light of the subsequent performance of the Japanese economy.

said, had led many companies to cut back capital spending and to take on more debt, in order to boost dividends and increase share buybacks. He urged CEOs to focus more strongly on long-term growth strategies.[6] A US academic, William Lazonick, has argued that share buybacks, partly influenced by the desire on the part of corporate managers to boost their share-based remuneration, have led to under-investment in innovation, in pharmaceuticals and in other industries.[7]

By contrast, another academic, Mark Roe, believes that concern over short-termism is exaggerated. 'Institutional investors', he writes, 'have not penalised companies for bumping up research and development spending even though the payoff may be years down the road. . . . High-stock valuations in favoured sectors—dot-com in the past, high-tech and biotech today—represent stock market *long-termism*, as many of these companies have no hope of producing revenue or profits in the short term to justify their lofty valuations.'[8] Roe accepts that some share buybacks are done for the wrong reasons, but suggests that in many cases companies are right to return surplus cash to shareholders rather than spend it on low-return investments.

There continues to be controversy in the US over short-termism, and over the role of activist investors in pressing companies to put more emphasis on maximizing shareholder value.[9] However, there is not much doubt that the US financial system is exceptionally well equipped to support young firms in emerging, high-technology industries. Capital for such firms, whether from business angels, venture capitalists and other private investors, or the stock market, is more readily available in the US than in other countries. As discussed in Chapter 2, this is one of the factors that has underpinned the success of biotech in the US.

Among European countries the UK has a financial system which is closest to that of the US. It has more public companies than, for example, Germany, and most of these companies are widely held, exposing them to the threat of takeover if their performance deteriorates. The UK also has a relatively large

---

[6] Letter sent by Larry Fink, chairman of BlackRock, to chairmen and CEOs, 21 March 2014.

[7] William Lazonick, 'Profits without prosperity: the problem with stock buybacks: how maximising shareholder value stops innovation', *Harvard Business Review* (September 2014); William Lazonick and Öner Tulum, 'US biopharmaceutical finance and the sustainability of the biotech business model', *Research Policy* 40 (2011) pp. 1170–87.

[8] Mark Roe, 'The imaginary problem of corporate short-termism', *Wall Street Journal* (17 August 2015), and by the same author, 'Share buybacks are not the problem', *Financial Times* (7 September 2015). See also Mark Roe, 'Corporate short-termism—in the boardroom and the courtroom', *Business Lawyer* 68, 4 (August 2013) pp. 977–1006.

[9] For a careful review of the evidence, which concludes that hedge fund activism does shorten the investment horizon of corporate managers, see John C. Coffee and Darius Palia, 'The wolf at the door: the impact of hedge fund activism on corporate governance', Columbia Law School, Working Paper 521 (New York 4 September 2015). For a defence of shareholder activism, see 'Capitalism's unlikely heroes: why activist investors are good for the public company', *The Economist* (7 February 2015).

Table 7.1. European and US publicly listed biotechs at the end of 2013

Country	No of public companies	Market capitalization ($m)
US	339	632,975
UK	30	32,825
Sweden	24	9,451
France	23	11,532
Germany	13	3,469
Norway	9	3,070
Denmark	9	15,766
Switzerland	8	10,614
Belgium	6	2,483
Netherlands	3	5,813

Note: It should be noted that E&Y's definition of biotech in the data supporting this table includes therapeutic and non-therapeutic firms, in contrast to the focus in this book. Nonetheless the E&Y data are illustrative of the scale of the UK industry compared to other European countries.

Source: Ernst & Young Beyond Borders 2014.

venture capital industry and, thanks to the changes instituted by the London Stock Exchange in the 1990s (described in Chapter 3), a more receptive stock market for early-stage firms than most other European countries; this is reflected in the number and market value of the UK's publicly listed biotech firms (Table 7.1). Yet the UK has been much less successful than the US in converting promising small firms in high-technology industries into medium-sized or large ones.

Is this due to flaws in the British financial system, or to other factors?

The Changing Funding Environment

In biotech, the British financial system has been criticized on two main grounds. First, biotech firms are said to have had less funding relative to their US counterparts, both in their early phase, when they rely on venture capitalists and other private investors, and when they go public.[10] Second, too many British firms are said to have been sold too early, before they have reached their full potential.

As the Spinks report noted in 1980, there was less venture capital available at that time for biotech in the UK than in the US, but the number and size of British venture capital firms increased over the subsequent decade. A further stimulus for the venture capital industry came in the early 1990s, when the

[10] William Bains, 'What you give is what you get: investment in European Biotechnology', *Journal of Commercial Biotechnology* 12, 4 (2006) pp. 274–83; Graham Smith, Muhammad Safwan Akram, Keith Redpath, and William Bains, 'Wasting cash—the decline of the British biotech sector', *Nature Biotechnology* 27, 6 (2009) pp. 531–7.

London Stock Exchange changed its rules to allow loss-making biotech firms to obtain a listing; this was followed in 1995 by the introduction of AIM, which enabled smaller firms to attract funds from public market investors. As in the US, venture capital firms were supported by institutional investors such as pension funds and insurance companies, some of which also invested directly in biotech companies.

Several of the first generation UK biotechs raised substantial sums before and after they went public. British Biotech, for example, raised just over £63m in four private placements before its IPO; most of the money came from UK institutions, but the company also attracted investors from the US and Japan. The flotation in 1992 raised a further £27.6m, and it was followed by several follow-on capital issues. Another well-financed firm was Oxford GlycoSciences, which raised nearly £50m between its foundation in 1988 and its flotation in 1997; the IPO brought in £31m and OGS raised £190m in two further capital issues in 2000. Neither of these firms did as well as their investors had hoped.

For firms that wanted smaller amounts of money, AIM was an alternative to venture capital and could be used as a preparatory step towards a listing on the main market. Oxford BioMedica, for example, raised £5m from its IPO on AIM in 1996, followed by four further issues between 1998 and 2000, raising £24m in total. In 2001 the firm moved to the main market, raising £36m.[11]

In this first phase, when the UK biotech sector was taking shape, most of the new firms were able to attract the capital they needed. Indeed, the majority of biotech drug developers that were founded before 1995 obtained a stock market listing, although, as subsequent experience was to show, the new-found enthusiasm for biotech firms was at the expense of selectivity.

As discussed in Chapter 4, there was a negative shift in investor attitudes in the early 2000s. Confidence was undermined by a series of company failures, bringing home to investors that biotech was a high-risk business and that only a tiny proportion of drugs that enter clinical trials reach the market. If the failures had been offset by some clear successes, as had happened in the US, investors might have continued to support the sector. A few continued to invest, and they were rewarded when some of the companies in which they held shares, such as Celltech and PowderJect, were sold. But the sector as a whole was out of favour in the stock market.

Venture capital firms also retreated from biotech; for established firms such as 3i this was part of a wider shift away from investing in all high-risk, technology-based firms towards management buy-outs. A striking difference

---

[11] Andrew Wood, CFO of Oxford BioMedica, 'Direct costs of share issues on public markets—reviving the debate over non-pre-emptive share issues', BioIndustry Association, 16 December 2012.

between venture capital in the US and UK, apart from the greater size of the American industry[12] and its higher appetite for high technology,[13] is the extent to which US firms have committed a much larger proportion of their funds to early-stage ventures; they have done so because these investments generated high returns, which was not the case in the UK or Continental Europe.[14]

As for the stock market, apart from a handful of flotations between 2004 and 2006, the IPO window on the LSE's main market was closed for most of the period between 2001 and 2014, although AIM continued to be open to smaller firms. Twenty-two drug discovery firms listed on AIM between 2000 and 2007—and some of them served their investors well. However, few of them made the transition to the main market, and the number of AIM-listed biotech firms fell sharply after the financial crisis.[15] It was not until 2014, when investor sentiment towards the sector was improving, that AIM again began to attract a significant number of drug discovery IPOs.

It was not biotech as such, but UK biotech, which was shunned by investors. There was money to be made in US biotech and that was where the attention of investors was focused. For example, the Reabourne Merlin Life Sciences Investment Trust, which had been set up in 1997 to invest in British and European firms, was reorganized and refocused in 2005. Management of the trust, which was renamed the Biotech Growth Trust, was transferred to OrbiMed, a leading US-based healthcare fund manager, and the portfolio was switched almost entirely to American firms, mainly NASDAQ-listed but early-stage, and not yet making a profit. Most of the investors in the trust remained British.

The shift to the US reflected the fact that the American biotech sector had produced a series of outstanding successes both in terms of blockbuster products launched and companies sold, making shareholders a great deal of money. There was also a view among some British investors that the quality of management was higher in the US than in the UK. Antony Milford of Framlington Investment Managers, one of the most successful healthcare investors, told a House of Commons committee in 2002: 'A lot of the people I meet running the American biotech companies are of the highest quality

---

[12] As a whole the US VC industry invested $20bn in 2010, while the UK invested only £1bn, making the US industry nearly three times larger as a proportion of the US economy than the UK industry is to the UK economy (at 0.14% of GDP vs. 0.05% respectively). Josh Lerner, Yannis Pierrakis, Liam Collins, and Albert Bravo Biosca, 'Atlantic drift: venture capital performance in the UK and the US' (Nesta Research Report June 2011).

[13] Hannah E. Kettler and Steven Casper, *The Road to Sustainability in the UK and German Biotechnology Industries* (London: Office of Health Economics. 2000).

[14] Lerner et al., 'Atlantic drift'.

[15] Colin Haslam, Nick Tsitsianis, and Pauline Gleadle, 'UK bio-pharma: innovation, re-invention and capital at risk', Institute of Chartered Accountants of Scotland, 2011.

both as scientists and managers. I am sorry to say that I have not, even on a relative size of economy basis, met as many outstanding managers in the UK sector.'[16]

In view of the poor performance of some of the firms that had listed in London in the 1990s, it is not surprising that investors had doubts about the quality of management in UK biotech. But did the shift from enthusiasm to pessimism go too far? Was the dearth of large, self-sustaining biotech firms in the UK after 2000 due to excessive caution on the part of investors, so that even the most promising and best-managed firms were starved of the capital they needed?

Underfunding

Biotech firms go out of business for a variety of reasons, of which the most common is the failure of their lead drug candidate in clinical trials. While well-funded firms are more likely to survive such setbacks than poorly funded ones, it is often difficult in any individual case to find a direct causal link between the declining fortunes of a company and its lack of access to finance. Several of the British firms that have failed over the last decade had raised substantial sums from investors since their foundation. Antisoma, which had listed in London in 1998 and went out of business in 2011, had raised £161m in equity funding since inception. Ark Therapeutics, which collapsed in 2010, had raised £158m. Another potential star, Renovo, had also been well supported by investors; it had raised just under £100m before its lead anti-scarring drug failed in clinical trials, leading to the extinction of the business.

These were substantial sums by European standards, but the amounts of capital available to their US counterparts was larger. 'A major difference in transatlantic investor attitudes,' Ernst & Young wrote in its 2001 report on European biotechnology, 'is that European investors seem more inclined to keep companies on a short leash and drip feed the funds. US investors seem much more willing to provide the hundreds of millions of dollars needed to grow companies aggressively into jumbo-sized packages.'[17]

The report compared the ten most valuable US companies that floated in 1996 with the corresponding group of European companies which had IPOs in the same year. By the end of 2000 the American group had an aggregate market capitalization of about €20bn and had raised €4bn in follow-ons and

---

[16] Anthony Milford, head of the healthcare and investment team at Framlington Investment Managers, evidence to House of Commons Select Committee on Trade and Industry, 17 December 2002.

[17] Integration: Ernst & Young's Eighth Annual European Life Sciences Report 2001, pp. 21–3.

other post-IPO financings. By contrast the European group from the class of 1996 had a market capitalization of €12bn and had raised only €333m since IPO. Some British firms, including CAT, raised large sums during the 2000 boom, but, as Ernst & Young pointed out, the €153m raised by CAT did not compare with the sums raised by its two US antibody rivals, Abgenix and Medarex, which raised €556m and €440m respectively.[18]

Writing in 2006, William Bains showed that European companies received less funding than US ones at the crucial early stages of the company's development.[19] In his view underfunding was the primary reason for the weakness of European biotech, and he blamed this in part on the timidity of European investors. US investors who created firms such as Amgen and Genentech, Bains wrote, believed that by boldly investing in world class science and management they could build world-class companies. European investors for the most part lacked such a vision.

If there was timidity on the part of investors, it was partly due to the 2000–01 boom-and-bust in biotechnology shares, which had a chilling effect on investor attitudes throughout Europe. The extreme case was Germany, where, as discussed in Chapter 6, the collapse of the Neuer Markt exposed the frailty of many of the biotech firms that had come to the market and damaged the reputation of the sector as a whole. German biotech has continued to suffer from a lack of support from public market investors and increasingly also from venture capitalists; since the Wilex flotation in 2006 there have been no biotech IPOs on the Frankfurt stock exchange.

The UK, too, had its share of biotech disasters from the late 1990s onwards. These events made investors realize how difficult it was to distinguish good firms from bad ones. If they were interested in high-risk, high-reward investments, they were inclined to switch to other industries, such as mining or oil and gas.

There was also a view, from specialist healthcare investors, that the quality of many of the firms looking for capital in the early 2000s was poor. 'The idea that the industry has been held back by lack of funding is wrong', one fund manager said. 'The crux of the matter was that it was not worth investing in these assets—they were not good enough, not innovative enough, they did not have management teams that had done it before—there was not the ground-breaking, exciting stuff that ultimately wins out.'[20] Most of the UK biotech firms coming to IPO were focusing on traditional small molecule

---

[18] Like CAT, both these firms were later acquired by larger companies but at a much higher price. Abgenix was bought by Amgen in 2005 for $2.6bn, Medarex by Bristol Myers Squibb in 2009 for $2.2bn. This compares with the $1.1bn paid by AstraZeneca when it bought CAT in 2006.
[19] William Bains, 'What you give is what you get', pp. 274–83. See also William Bains, *Venture Capital and the European Biotechnology Industry* (London: Palgrave Macmillan, 2009) pp. 73–85.
[20] Interview with Gareth Powell, Polar Capital, 14 October 2013.

drugs and/or methods of drug delivery, and not therapies based on novel technologies such as monoclonal antibodies, recombinant proteins, or gene therapy (see Figure 3.1 in Chapter 3).

Thus there are several possible explanations for the retreat of UK investors after the failures of the late 1990s and early 2000s: excessive caution, a rational response to earlier disappointments, and doubts about the quality of the firms that were seeking capital. Yet in comparing the UK and the US there was a crucial difference in the way the two biotech sectors had started and evolved, and this difference has to be taken into account in considering their access to finance.

In 1980, when Celltech was formed and Genentech had its IPO, the US already had a well-developed venture capital industry and, in NASDAQ, a stock market that was receptive to early-stage high-technology firms. The leading venture capital firms, some of which were founded and run by people with industrial rather than purely financial experience (Tom Perkins at Kleiner Perkins was an outstanding example), had mentoring skills that were less well developed in the UK. In all these areas the UK was a late starter, and this lack of experience contributed to some bad investment decisions. When the London stock market opened up in the 1990s, there were too few investors with the knowledge and expertise to distinguish between strong and weak firms.

In 1993, when the wave of biotech IPOs in London was getting started, well over fifty US biotech firms were already listed, and some of them, including Genentech and Amgen, had already launched big-selling drugs such as recombinant insulin for diabetes; these were replacements for known proteins and were sometimes described as low-hanging fruit. The success of these drugs, launched within a few years of the companies' IPO, gave US investors confidence in the sector and made them more willing to support later entrants using novel technologies such as monoclonal antibodies. UK biotech firms missed out on the low-hanging fruit (as did their rivals in other European countries). Those that attempted to commercialize novel technologies, such as Celltech and CAT with monoclonal antibody drugs, faced much greater difficulty and delay in bringing drugs to the market than US firms had done with recombinant protein drugs.[21]

Thus the differences in investor attitudes and behaviour between the US and the UK in the early years cannot be separated from the fact that one of the two biotech sectors started earlier than the other, faced a different set of opportunities, and benefited from the existence of institutions—NASDAQ and venture capital—which were already in place. While it was possible to create comparable institutions in the UK, that was not enough to allow British

[21] J. M. Reichert, C. J. Rosenweig, L. B. Faden, and M. C. Dewtiz, 'Monoclonal antibody successes in the clinic', *Nature Biotechnology* 23, 9 (2005) pp. 1073–8.

firms to catch up. In the US success built on success, creating a momentum of growth which, despite the ups and downs of the stock market, encouraged investors not only to continue supporting the sector but also to devote considerable resources, in the form of well-qualified analysts and scientists, to understanding it. In the UK there were too few successes, and too many failures.

Were the Best Firms Sold Too Early?

The view that too many promising British biotech firms have been sold too early has been a recurring theme in the numerous government-commissioned reports that have been published since 2000. The authors of these reports recognized that a sale to Big Pharma, or to another biotech firm, was an entirely rational move for many small biotechs, and often preferable to continued independence. This applies as much to the US as to the UK. A trade sale can be the best way of maximizing value for the company's owners and of ensuring that its technology is put to good use. But it has also been argued—for example in the 2003 BIGT report—that the sector needs beacons or bellwethers which can serve as role models for investors and for would-be company builders. The UK biotech sector would be healthier, according to this view, if some British firms had stayed independent for longer, growing into profitable, mid-sized firms that come to the stock market and retain the confidence of their investors for a long period.

Of the companies which looked capable of becoming beacons, Celltech and Cambridge Antibody Technology in particular have been regarded as missed opportunities. In the case of Celltech, which was bought by UCB in 2004 (after Celltech had been in existence for 24 years), the decision to sell followed an internal debate on how best to exploit what had become the company's most valuable asset, the rheumatoid arthritis drug known as CDP870 (later given the trade name Cimzia). The choice was between developing the drug on its own, which would have required a large injection of new capital, and seeking a partner or finding an acquirer for the company as a whole.

In view of stock market conditions at the time there was little prospect of raising the necessary amount of capital in the London market, and Celltech was not attractive enough to US investors to permit a capital-raising exercise in New York. By selling to UCB at a price of £1.5bn Celltech was giving its investors a good return and ensuring that its best asset would pass into the hands of a company capable of taking it through to the market.

In an interview some years later Peter Fellner, Celltech's chairman, said that the company might have stayed independent for longer if it had taken advantage of the 2000 boom to use its highly valued shares to raise more capital and to make another large revenue-producing acquisition, comparable to the Medeva deal in 1999. 'I wanted to do it and we had several prospects but the Board rejected the idea. If we had done those two things our financial position would have been stronger, but that was our fault, not the fault of the financial system. We did not manage our financial engineering well enough.'[22]

Cambridge Antibody Technology is often seen as even more of a disappointment because of its role in the development of Humira (adalimumab), the blockbuster rheumatoid arthritis treatment discussed in Chapter 4. The origins of Humira lay in a licensing agreement which CAT made with Knoll, BASF's pharmaceutical subsidiary, in 1993. It was one of several partnerships which CAT negotiated during the 1990s (when it was still a private company) as a means of generating the revenue needed to maintain its research programmes. No one anticipated that the BASF project would produce the world's highest selling drug.

At the time of the licensing deal with BASF the principal shareholder in CAT was Peptech, an Australian company which had seed-funded the company at the start. CAT's financial position, and its ability to strike bargains with licensees, might have been stronger if more venture capital firms had taken a stake in the company at an earlier stage. (3i came in later, shortly before the IPO.) They might also have injected more commercial drive into a company that was stronger in early-stage research than in development. However, at that time venture capitalists were reluctant to invest in what was still an unproven technology. It was not until the IPO in 1997 that CAT acquired the financial resources that were needed to support its chosen strategy, which was to supplement its licensing deals with a programme of in-house drug development.

The fact that most of the profits earned by Humira went to an American rather than a British company (Abbott of the US having bought BASF's pharmaceutical division in 2000) has been seen as another example of innovative British science being exploited by non-British companies. It is certainly arguable that CAT could have played its cards more skilfully. Its choice of drug programmes to pursue in-house after the IPO was lacklustre. It was involved in drawn out and distracting legal disputes, with Abbott in the US and Morphosys in Germany, and these disputes cast a cloud over CAT's stock price. But even if these hazards had been avoided, CAT was likely, sooner or later, to have become a takeover candidate. By the early 2000s most of the big pharmaceutical companies had recognized the importance of monoclonal antibodies and were looking for acquisitions that would give them an entrée into this field. Not long after the CAT/AstraZeneca deal, Domantis, Greg Winter's second antibody company, was bought by GlaxoSmithKline. In the

---

[22] Interview with Peter Fellner, 5 June 2013.

case of both CAT and Domantis, their value lay in their technology platforms, and the two British companies that acquired them planned to use these platforms to strengthen their own pipelines.

Celltech and CAT were listed companies. Those that were not yet public were in a different situation. Most of them were owned by venture capital firms, which attract money from investors—their limited partners—by creating funds that have a fixed term, normally 10 years; the limited partners expect to see a return on their money within that 10-year period, although an extension can sometimes be negotiated. This has two consequences. One is that when the fund has been in existence for 5 or 6 years the general partners who are running it start looking for ways of exiting their investments, either through a trade sale or an IPO. The other is that the venture capital firms which have invested in the company may have done so at different stages in their fund's life; those whose funds are nearing the end of the 10-year term may be more eager to secure an exit than firms whose funds are at earlier stage. These issues do not arise in the case of evergreen funds—for example, the corporate venture capital arms of Big Pharma or the Wellcome Trust's Syncona—which are not constrained by exit deadlines, although these funds often invest as members of a consortium that includes conventional venture capital firms.

Decisions on when and how to exit an investment are influenced partly by the internal financial situation of the venture capitalist owners, partly by the view they take of how best to maximize the value of the investment. In some cases, like that of Arrow discussed in Chapter 4, the need for an early exit may conflict with what the founders, senior managers, or even some other investors may think is in their own and the company's best long-term interests. But such disagreements are inherent in the venture capital business model, and they can occur in the US as well as in the UK.

Judgements about whether any particular sale has come too early depend on highly uncertain forecasts about how quickly the company's drug candidates might get through clinical trials and reach the market—if indeed they ever do so. A Big Pharma buyer may take a more optimistic view of the target company's prospects than its owners and be prepared to pay a high price—and it may make mistakes.[23] There are numerous cases where purchases by Big Pharma have turned out badly because the acquired firm's candidates failed in clinical trials; in these cases, whether through luck or shrewd judgement, the venture capitalists did better by selling out than they would have done if they had continued to plough money into the business.

---

[23] In 2008 GSK bought Sirtris, a Boson-based biotech firm for $720m, but the technology that Sirtris had been developing did not generate the results that the acquirer had hoped for. The business closed down in 2013.

As for the choice between a trade sale and an IPO, this depends on how far the company's drug development programmes have advanced and on whether, when the owners are looking for an exit, the stock market is receptive to biotech IPOs. In the UK, for most of the period between 2001 and 2014, there has been little scope for IPOs on the LSE's main market. This has inevitably led to a greater reliance on trade sales.

The pressure for early exits has often been seen, by founders of companies and by policy makers, as a weakness in the venture capital model. (See the Avidex case discussed in Chapter 5.) The need for a different approach to biotech investment was part of the rationale for the Wellcome Trust's decision to set up Syncona in 2012. That decision was welcomed in the beleaguered biotech community, but what would have been even better, from the point of view of biotech firms, was a change of attitude on the part of the big investing institutions—not only a greater willingness to invest in the sector, but also a willingness to align their investment approach to the long-term nature of the drug discovery business.

If there had been a group of patient, long-term shareholders, with deep pockets, backing several companies like Cambridge Antibody Technology, investing in them while they were still private and providing them with regular injections of capital through to the IPO and beyond, might one or more of them have emerged as a world leader in a novel technology, performing the beacon role that was so badly needed to raise the profile of the sector in the City, and in the world at large?

It is the behaviour of institutional investors, rather than that of venture capitalists, which has been central to the debate about short-termism in the UK. According to some observers, the preoccupation of financial institutions with the short-term performance of the shares in their portfolio has damaged the biotech sector, contributing both to the underfunding of biotech firms and to premature sales. An influential figure in this debate has been Neil Woodford, who was formerly a fund manager with Invesco but since 2014 has been running his own investment firm.

A Patient, Long-term Investor

Woodford's involvement with biotech began in the mid-1990s when he was an investment manager in the unit trust group, Perpetual Investment Management. This firm was a large shareholder in British Biotech and several other pioneering British biotechnology firms.[24] When Perpetual merged with

[24] Perpetual held just under 10 per cent of British Biotech at the time of that company's crisis in 1998.

Amvescap in 2002—the merged firm was called Invesco Perpetual—he was put in charge of UK equities and he acquired a reputation as one of the City's most astute fund managers.

Woodford's success at Invesco stemmed in part from good timing—buying tobacco shares in the late 1990s, for example, when the stock market was infatuated with internet firms—and the bulk of his fund's investments were in well-established companies, including pharmaceutical companies, which generated a reliable flow of dividends. But he devoted a small part of his portfolio—around 5 per cent—to small science-based firms, principally in biotech, which he believed were under-valued and capable in time of becoming highly profitable.

What was distinctive about his approach to these firms was his willingness to provide them with capital on a regular basis and to hold on to his shares for a decade or more before selling out. He was also unusual among UK-based institutions in allocating part of his portfolio to unquoted firms. As he explained, 'other investors gave up on biotech but I held on. There are lots of excellent prospects, especially in the unquoted sector, but you have to take a long-term view, wait at least 5–10 years, not 5–6 weeks or months. Other investors have disqualified themselves from the sector because they don't have the right investment horizon or the right attitude to risk.'[25]

Woodford encourages firms to stay unquoted and only float when they are mature and self-sustaining from a cash flow point of view. 'If they float too early there are a lot of burdens in being public with a diversified shareholder base. If they don't deliver in the short term or have setbacks the share price collapses and they tend to become another orphan asset, another living dead. Shareholders give up and they can't raise capital.'

In Woodford's view the UK has many of the ingredients necessary for a flourishing biotech sector—most importantly outstanding science—but it suffers from the short-termism that is endemic in the asset management industry. 'My industry has a lot to answer for. Our job is to take savings and invest in businesses that need capital. But there are too many intermediaries, pension consultants and so on, getting in the way. Fund managers are obsessed by quarterly performance, they are poorly equipped to do their proper job.'

Some of Woodford's investments in biotech were made through Imperial Innovations and IP Group. Invesco acquired a large shareholding in both these organizations, and it also invested directly in some of their portfolio firms.[26] What he liked about them was their deep pool of expertise in

---

[25] Interview with Neil Woodford, 6 May 2013.

[26] Invesco was also a large shareholder in another technology commercialization company, Netscientific, which was floated on AIM in 2013.

university science and their ability to nurture young firms outside the glare of public markets. Woodford did not impose exit deadlines on the firms in which he invested. 'We don't sit down with companies and say here is our invest-ment, when will we get our money back?'

Few other UK-based institutional investors shared Woodford's approach. One that did so was Lansdowne Partners, a hedge fund which invested alongside Invesco in Imperial Innovations and IP Group, and in several biotech firms.[27] Peter Davies, senior partner, was as convinced as Woodford that the commercial potential of UK biotech had been under-exploited, and he believed that the best way of exploiting it was through aggregators like Imperial Innovations and IP Group. He drew a comparison between what these two organizations were doing and what 3i had done in the 1980s and 1990s. 3i used the capital provided by its owners—the banks that had set up ICFC after the war—to make loans to, and to take equity stakes in, a range of small and medium sized firms around the country. 'The key thing', Davies said, 'is to have permanent or semi-permanent capital. Neither the venture capitalists nor AIM can provide that, but Imperial Innovations and IP Group can.'[28]

Other fund managers, while acknowledging Woodford's record of success and praising his contribution to the biotech sector, questioned whether the allocation of a small but significant amount of the funds entrusted to him by his investors to high risk biotech firms was justified by the likely returns. They also pointed out that the size of his funds gave him a freedom which other fund managers lacked. The failure of one or more of the firms which he backed would make no more than a marginal difference to the performance of the fund as a whole. A smaller fund, especially one that specialized in healthcare and might have a larger proportion of its investments in biotechnology, could not afford to take those sorts of risks.

For most fund managers the idea of investing in a loss-making biotech firm which might not achieve a return for 10 or 15 years was not attractive. As Daniel Mahony of Polar Capital, a London-based healthcare investor, remarked, 'I am not sure that many people would regard a 5–6 year time horizon as short term. If I made an investment and it delivered little to no investment return for six years, I am not sure if my shareholders would be particularly pleased. I think most investors have been avoiding the UK biotech

---

[27] In 2015 the principal shareholders in Imperial Innovations, which was listed on AIM, were Invesco (42.0%), Imperial College (20.1%), Woodford Asset Management (14.2%), and Lansdowne Developed Markets Master Fund (13.7%). IP Group was listed on the main market, with Invesco holding 25.7%, Lansdowne Partners 15.1%, Baillie Gifford 10.5%, and Woodford Investment Management 7.1%.

[28] Interview with Peter Davies, Lansdowne Partners, 6 May 2013.

sector because they can't see the long term potential—therefore there is no short term potential either.'[29]

The UK's leading financial institutions, such as Legal & General, Aviva, Standard Life, and Baillie Gifford, are substantial investors in British health-care companies, but most of them have tended to avoid high-risk drug dis-covery firms. There are also a few UK-based specialist healthcare funds, but most of their biotech investments have been in the US. In 2014 the Biotech Growth Trust, managed by Orbimed, had investments in forty-six firms, only one of which, GW Pharmaceuticals, was British. In that year London-based Polar Capital set up a new Biotechnology Fund, which had 76 per cent of its investments in the US, and only 1 per cent in the UK.

The Enduring Appeal of NASDAQ

Woodford's approach is more likely to be followed by other institutions if some of the firms in his portfolio grow into large or at least medium-sized businesses, listed on the stock market (preferably in London rather than New York) and continuing to retain the support of investors through outstanding financial performance. That was why the Circassia flotation in 2014, discussed in Chapter 5, was so important, both for Woodford's reputation and for the biotech sector as a whole. If the London Stock Exchange could re-establish itself as an attractive venue for biotech IPOs, with institutional investors supporting the newly listed firms, this would fill a gap in the UK's biotech ecosystem, give venture capitalists an alternative exit route to the trade sale, and make it less likely that promising British firms would emigrate to the US.

Yet at the time of the Circassia flotation the magnetic pull of NASDAQ remained strong. In 2014 and 2015 two AIM-listed companies, GW Pharma and Summit, listed their shares on NASDAQ and subsequently raised capital in the US. More significant was the decision by Adaptimmune, a highly regarded unlisted firm that had been supported for several years by private investors in the UK but had recently attracted investment from the US, to launch its IPO in New York rather than London.

As a venue for biotech firms NASDAQ was far larger than any non-American stock exchange, and it had attracted a wide range of investors who invested in the sector in different ways and at different times. They included specialist healthcare investors with knowledge of the sector as well as large generalist investors such as Fidelity. The ability to assess and value biotech firms was far more highly developed than in London, where there was only a handful of

---

[29] Interview with Daniel Mahony, Polar Capital, 9 May 2013.

specialist investors and not many analysts in the big institutions who were deeply immersed in the sector.

As one former analyst recalled, 'When I was a drug analyst in the City I generally felt I could hold my own on this side of the Atlantic when it came to the gritty details of drug R & D. However, that feeling usually evaporated when I was working in Boston. Working in Boston felt a bit like the transition from primary school to big school. It is hardly an exaggeration to say that I could meet more expert health care investors in Boston in a day than I could meet in London in a year.'[30] Adam Kostyl, head of European listings at NASDAQ, has another reason for why firms listed in Europe rather than in the US are at a disadvantage: 'The problem in Europe is not the day of the IPO, it is the lack of ongoing support in terms of liquidity, follow-on fundraising and analyst coverage.'[31]

The dearth of specialist healthcare investors in London meant that firms considering an IPO in London had to rely mainly on generalists. Philip Astley-Sparke, formerly chief financial officer at Biovex, and later a partner at Forbion, the Dutch venture capital firm, said that, historically, biotech listings on the London Stock Exchange did not take place unless generalists were involved to a large degree. 'In the States no generalists are required. These UK generalists are unlikely to be diversifying into US biotech. Hence, a few UK biotech IPOs may get done and then a single disappointment sends the generalists running for cover. By contrast, a few blow-ups on NASDAQ is just noise.'[32]

Compared to the US, the investor pool in London lacked depth and specialization. This difference in scale and expertise gave NASDAQ a decisive edge over the London Stock Exchange. US institutional investors were not necessarily more long-termist in their investment approach than their British counterparts, although some of them were prepared to hold large stakes in favoured companies for a number of years, as long as those companies performed well. For example, Fidelity invested in Alnylam (a company discussed in Chapter 2) at the time of the IPO in 2004 and continued to hold 15 per cent of the shares in 2015; such investors keep in close touch with the management and are well informed about the progress of the company's research. But companies that perform badly cannot count on the loyalty of shareholders; they put themselves at risk of being attacked by activist investors who might demand changes in strategy or even the sale of the business. One of the best known activist investors, Carl Icahn, targeted several large and well-established

---

[30] Interview with Jack Scannell, former senior analyst with Bernstein Research, 23 March 2015.

[31] Andrew Ward 'European biotechs rush to list on Nasdaq', *Financial Times*, 15 July 2015.

[32] Quoted in Steve Dickman, Boston Biotech Watch, 'Why have so few European biotechs made it through the IPO window? http://bostonbiotechwatch.com/2014/10/23/why-have-only-a-few-european-biotechs-made-it-through-the-ipo-window/.

biotechnology companies when they ran into financial difficulties; two of them, MedImmune and Genzyme, were subsequently taken over.

It is sometimes said that American investors, both private and institutional, have a greater appetite for risk than their European counterparts; they are willing to take bigger risks in the hope of higher rewards. A New York analyst described what he saw as an important difference between the two sets of investors. 'I frequently hear US biotech investors saying that they would never sell the stock at a 25 per cent premium over what they paid for it—they would expect to hold it until it doubles or triples—whereas the European generalist investor is happy to sell for the 25 per cent premium first offer from the opportunistic pharmaceutical company.'[33]

This difference could be attributable not to cultural factors, but to the greater knowledge and experience acquired by the US biotech investor, enabling him or her to make an informed judgement about the long-term prospects of the company concerned. The extent of an investor's risk tolerance depends to a large extent on his or her expertise in the sector, and the size and diversification of the fund. It is also affected by which market the investor is active in. For example, there are likely to be more exit options available in the US than in the UK.

Given NASDAQ's early start as a market for high-technology firms and the growth of the US biotech sector in the 1980s and 1990s, it was bound to be difficult for the London Stock Exchange to catch up. In theory a European equivalent to NASDAQ that brought together high-growth firms from all over the Continent, as advocated by Ronald Cohen and others in the mid-1990s, might have become a credible competitor to NASDAQ.[34] However, EASDAQ, the new market which Cohen and other venture capitalists established, was never able to attract sufficient support either from European investors or from firms planning an IPO; the few biotech firms which briefly listed on EASDAQ, such as Antisoma in the UK, switched back to their local exchanges. The political and practical obstacles to a unification of European stock exchanges were, and have remained, insuperable.

The Role of AIM

AIM, launched in 1995, was intended for small firms in general, not specifically for high-technology firms.[35] Between 1995 and 2015 some 3,000 firms

[33] Geoff Porges, Bernstein Research, private communication.

[34] Dana T. Ackerly, 'Easdaq—the European market for the next hundred years?', *Journal of International Banking Law* 12, 3 (1987) pp. 86–91.

[35] Sridhar Arcot, Julia Black, and Geoffrey Owen, *From local to global, the role of AIM as a stock market for growing companies* (London Stock Exchange, September 2007).

have listed on AIM, raising a total of over £90bn. Taken in aggregate, the performance of AIM-listed shares during this period has been poor.[36] The market has also been criticized, not least in the US, for failing to impose adequate governance standards on its listed firms. Yet some AIM companies have performed outstandingly well, and within the biotech community the market has some strong defenders.

David Evans, an entrepreneur who has been chairman of several AIM-listed diagnostics and biotechnology firms, including Epistem and Scancell, says that AIM should be recognized for what it is, a listed alternative to venture capital. 'AIM is a more efficient way of raising money than venture capital. From a management perspective managers are accountable for their actions and have the freedom to do the things that are needed in the interests of the company, whereas with VCs on your board you are always having to manage not just them but also their inherent conflict of interest which many are oblivious to. AIM for me has often succeeded where venture capital has failed, notably with Scancell and Epistem. Not only were we able to raise sufficient capital, but that capital has been deployed in a manner reflecting the growth in value over the period.'[37]

Others are more critical, pointing out that shares in pure drug discovery firms may move violently up or down in response to good or bad news; these firms are also generally too small to attract analyst coverage. For GW Pharmaceuticals, whose move to NASDAQ was described in Chapter 5, the IPO on AIM in 2000 provided an entrée into the public market and made investors aware of its existence, but the subsequent performance of the shares was erratic. Although it had a loyal long-term institutional investor in the M & G Recovery Fund, AIM did not provide the pool of well-informed healthcare investors which was available in the US. That was part of the reason for GW's decision to list on NASDAQ; it never seriously considered a move to the LSE's main market.

AIM investors generally prefer revenue-producing firms whose strategy they can understand. For example, Abcam, the Cambridge-based producer and distributor of reagents and antibodies, went to AIM in 2005 when it was already profitable, and subsequently became one of the most highly valued AIM shares. Among the biotechnology-related firms that listed on AIM in 2014 was another Cambridge firm, Horizon Discovery, a provider of technology and services to other companies. Not long after its IPO Horizon Discovery used its AIM-listed shares to make a large acquisition in the US.

---

[36] According to Paul Marsh of the London Business School £100 invested in AIM in 1995 would have become £83 in 2015, even with dividends reinvested. *Financial Times*, 19 June 2015.
[37] Interview with David Evans, 17 July 2013.

Institutional investors have mixed views about AIM, with some staying away on the grounds that there is too little liquidity in the shares, too many fragile companies, and too many cases of weak boards of directors. Others argue that, used properly, it can be a good training platform for ambitious companies. The M & G Recovery Fund, run by Tom Dobell, has about 10 per cent of its portfolio in AIM shares. 'AIM is not an end point', Dobell says, 'but it can be a useful proving ground—we want these firms to graduate to the main market.'[38]

Dobell's fund is one of the few generalist investors which has continued to invest selectively in UK biotech, through AIM as well as the main market. It has been a long-term shareholder in GW Pharmaceuticals, making its first investment in 2002; its patience was rewarded when GW graduated from AIM, not to the main market, but to NASDAQ. Dobell believes that AIM's rules could be usefully updated; a ban on short selling, he thinks, would redress the balance between short-term traders and long-term investors who are trying to provide capital for junior companies.

The Missing Tier in UK Biotech

While there are flaws in the UK financial system, as shown in John Kay's report on short-termism, it is difficult to establish a close link between those flaws and the dearth of large, British-owned biotech firms. Indeed, the UK has done relatively well in biotech compared to other countries which are generally considered to have more long-termist financial systems, such as Germany: Morphosys is the only German firm with a market capitalization in excess of $1bn. The fact that Actelion, Europe's most highly valued biotech company, is based in Switzerland is not because Switzerland has a more long-termist financial system than other European countries. The success of Actelion, described in Chapter 6, has more to do with company-specific factors rather than national ones; its starting point was a drug candidate acquired from Roche, rather than a novel technology platform coming out of an academic laboratory, and it was backed at the start by non-Swiss investors.

Disappointment in the UK's performance stems from the much greater success of the US in generating so many high-performing firms which brought valuable drugs to the market within a few years of their foundation. Out of that success came a supportive investment community which has continued to grow. The presence of so many investors with different and overlapping expertise (from early to late stage in the drug discovery process) means that

---

[38] Interview with Tom Dobell, 24 October 2013.

shares are more easily traded and no investors are forced to have a 15-year investment horizon. No European country has been able to create a biotech-focused investment community comparable to that of the US in terms of specialist knowledge, the amount of funds available, and the number of investors. The scale difference between the US investment community and its UK counterpart, and the greater appetite in the US for high-tech investments, has an important consequence: the UK is not simply a proportionately scaled-down version of the US sector.

The US has four big biotechs—that is, firms valued at over $50bn, comparable to some Big Pharma companies in market value. This is a small number relative to the sector as a whole, but the same is true in information technology; there are very few firms that can aspire to become Apples, Googles, or Amazons. In contrast the UK has no biotech firms of this size (see Table 7.2), and neither is there one in Europe as a whole. Given the scale of the US market, and the head-start of US biotech firms, it is not surprising that no European counterpart to an Amgen or a Gilead has yet emerged. Looking back to 2005 there was only one US biotech firm of this scale, Amgen (as shown in Table 7.2). The US advantage is less pronounced when considering firms in the $5bn–$50bn range, where the US has nine firms (four of which joined the stock market after 1997 and so are more comparable in age to their later emerging European counterparts). The UK has only produced one firm in this bracket—Shire, which is no longer headquartered in Britain—but then the UK's economy is only a fraction of the size of the USA's.

For the UK, or any other European country, to produce an equivalent to Amgen is unlikely in the extreme. A more realistic objective is the creation of a

Table 7.2. The US and UK therapeutic biotech sectors in 2005 and 2013: a size-based comparison of market capitalization, showing the percentage of firms in each size category (number of firms in brackets)

Market capitalization ($)	0–49.99m	50m–499m	500m–4.99bn	5bn–49.99bn	>50bn	Total number of firms
All US (2005)	12.5% (21)	64.9% (109)	18.5% (31)	3.6% (6)	0.6% (1)	168
US Post-97s (2005)	12.5% (11)	76.1% (67)	11.4% (10)	0	0	88
All UK (2005)	46.2% (12)	46.2% (12)	7.7% (2)	0	0	26
All US (2013)	19.5% (38)	42.6% (83)	31.3% (61)	4.6% (9)	2.1% (4)	195
US Post-97s (2013)	20.8% (31)	46.3% (69)	30.2% (45)	2.7% (4)	0	149
All UK (2013)	29.6% (8)	59.3% (16)	7.4% (2)	3.7% (1)	0	27

Note: Comparative data in Table 7.2 are based on a resource provided by Lähteenmäki and colleagues, from the annual audit published in *Nature Biotechnology* 'Public-Biotech: the numbers'. These articles have been published annually since 1997 (together with associated online data sheets) and provide a list of public firms in the biotech sector year by year. Since 2005 these data have contained information on market capitalization. Here we compare data from the earliest and most recent years available. Only firms focused on drug discovery and development are included. Additional data for UK firms have been added as necessary to ensure as full coverage as possible.

Source: *Nature Biotechnology* annual survey 'Public Biotech: the numbers'.

sufficiently large group of mid-sized firms—valued at, say, between $500m and $5bn—to provide the biotech sector with a degree of solidity and resilience. It is this middle tier, accounting for about 30 per cent of all US publicly listed therapeutics firms, that gives the US biotech sector its strength and endurance even as individual firms are acquired or decline (Table 7.2).The corresponding figure for the UK is only 7 per cent, reflecting the problems which UK biotechs have faced as they try to move from start-up and early-stage drug discovery to become sustainable, independent businesses.

The difference between the US and UK in the proportion of firms in the middle tier remains similar even considering only US firms with IPOs after 1997 (so counting only US firms joining the stock market after the boom in UK biotech shares in the mid-1990s). This shows that the large US middle tier is not just a product of the earlier emergence of public biotech firms in the US. Rather it is an indication that US firms are better resourced by their capital markets and become more highly capitalized more quickly than their UK counterparts.

This strength in depth is one of the factors that has distinguished the US from the British biotech sector over the past 20 years. The next chapter looks at another aspect of the environment in which British biotechs were operating: the role of public policy. Could successive governments have done more to improve the performance of British biotech firms?

8

The Role of Government

Don't doubt our ambition—not just to stay in the game but to lead the game. . . . I want the great discoveries of the next decade happening right here in British laboratories, the new technologies born in British start-ups, proven in British hospitals. David Cameron, British Prime Minister, December 2011[1]

There has been a long-running debate among economists and policy makers about what governments can do to promote new industries. One view is that governments should concentrate on creating an environment in which innovative activities of all sorts can flourish and leave it to the private sector to exploit whatever commercial opportunities may arise from scientific advances. Another is that some new industries have a special importance because of their growth potential, their contribution to exports and employment, or their wider role as suppliers of technology throughout the economy. In these cases the costs and uncertainties involved in commercial ventures may be too great to attract private sector investment on a sufficient scale, and the government has to step in to fill the gap.[2]

In the 1960s and 1970s the British and French governments used public money to support several of their high-technology industries. In computers, for example, both countries created national champions—International Computers Limited (ICL) in the UK and Compagnie Internationale pour l'Informatique (later merged with other firms to create CII-Honeywell-Bull) in France—in the hope that they would become strong enough to compete against IBM; both

---

[1] Speech at FT Global Pharmaceuticals and Biotechnology Conference, 6 December 2011.
[2] These two approaches to technology policy, one applying to all industries and the other specific to one particular industry, have been described as diffusion-oriented and mission-oriented, Henry Ergas, 'The importance of technology policy', in *Economic Policy and Technological Performance*, edited by Partha Dasgupta and Paul Stoneman (Cambridge: CUP, 1987). Others use the terms 'horizontal' and 'vertical' to distinguish the two approaches: Iciar Domingues Lacasa, Thomas Reiss, and Jacqueline Senker, 'Trends and gaps in biotechnology policies in European member states since 1994', *Science and Public Policy* 31, 5 (October 2004) pp. 385–95.

projects failed.[3] In the aircraft industry government intervention was more successful, thanks in part to the decision by the two governments, later joined by Germany, to switch from national support programmes to collaboration. The Airbus consortium became a credible competitor to Boeing, and is generally regarded as a success for European industrial policy, although the Airbus model, involving the concentration of resources in a single European company, has limited applicability to other industries.[4]

In the aircraft case, part of the rationale for government support was that capital markets were thought to be unable or unwilling to finance projects that might take 15 years or more to generate a return; without government intervention, the market for large civil airliners was likely to be dominated by the US. Some of the same considerations arose in biotechnology. US companies were the first to exploit the new techniques, and there were fears in several European countries, and in Japan, that without government support their biotech firms would be left behind. This was one of the arguments used by the Spinks Committee in the UK to support the case for creating a new, government-supported company in this field.

In the UK there were other reasons why governments took a special interest in biotech. British academic scientists had made an outstanding contribution to molecular biology, genetics, and immunology, the disciplines on which the new sector was based; this was a national asset which, if exploited effectively by British firms, could generate benefits for the economy. In addition, one of the principal applications of biotechnology was in medicine. While most other British industries had been losing ground in world markets since the end of the Second World War, British pharmaceutical companies were a shining exception. Given the close links between biotech and pharmaceuticals, it was reasonable to think that British firms could also be successful in this new field.

The success of the British pharmaceutical industry had been due, in part, to supportive government policies. Would the same policies be equally effective in biotech, or was something different needed? This chapter looks first at the role played by British governments in pharmaceuticals in the decades following the Second World War, before considering the impact of government on the biotech sector from the 1980s onwards.

---

[3] Geoffrey Owen, *From Empire to Europe, the Decline and Revival of British Industry since the Second World War* (London: HarperCollins, 1999) pp. 262–70.

[4] The civil airliner business is characterized by large economies of scale and scope, high barriers to entry, and predictable customer requirements. 'Designing a new aircraft is largely a matter of throwing money at the challenge of carrying a given number of passengers for a given distance and at reasonable speed and safety and at minimum fuel cost.' Paul Seabright, National and European champions—burden or blessing? *CESifo Forum*, 6, 2 (Summer 2005), published by Ifo Institute for Economic Research, Munich. See also Damien Neven and Paul Seabright, 'European industrial policy: the Airbus case', *Economic Policy* 21 (1995) pp. 313–58.

British Success in Pharmaceuticals

Before the Second World War British pharmaceutical companies were smaller and less committed to research than their principal competitors; the world leaders at that time were German and Swiss companies. The invention of penicillin and the developments which flowed from it after the war had a transforming effect on the British industry. Although penicillin was exploited more fully and more profitably in the US than in the UK, the pioneering work had been done by British scientists, and the success of the new drug encouraged British companies to step up their spending on research, both in antibiotics and in other therapeutic areas. By the 1980s the leading companies—Glaxo, Wellcome, Beecham, and Imperial Chemical Industries (ICI)—had emerged as significant players in the world pharmaceutical industry.

The most spectacular advance was made by Glaxo, thanks in large part to the success of Zantac, an anti-ulcerant drug which was launched in the UK in 1981 and the US in 1982. The marketing of Zantac in the US, involving first the acquisition of a small American pharmaceutical firm and then a co-marketing agreement with Roche, was the basis for Glaxo's emergence as one of the world's largest pharmaceutical companies. The principal architect of Glaxo's strategy during those years was Sir Paul Girolami, who was chief executive and later chairman from 1981 to 1994. A non-scientist with a background in finance, Girolami took the crucial decision to price Zantac at a substantial premium to the rival product, Tagamet from Smith Kline and French, on the grounds that it had therapeutic advantages. The higher price was accepted in the market, and the new drug quickly outsold its rival.[5]

Glaxo did not receive (as Celltech did in the 1980s) any direct government support, but British pharmaceutical firms were operating in a domestic environment which was influenced by government in two ways, through the regulatory system and through the use of public funds to support health-related scientific research.

As in the US, controls over the safety of drugs were tightened after the thalidomide tragedy in the late 1950s. The government established a Committee on Safety of Drugs, drawing on academic expertise and charged with the task of advising on the toxicity and efficacy of new drugs. The industry, through its trade association, agreed that no new drug would be introduced without the Committee's approval. This regulatory regime required manufacturers to demonstrate the efficacy of their drugs through well-conducted clinical trials. One consequence was to encourage firms to focus on innovative

---

[5] Edgar Jones, *The Business of Medicine, the Extraordinary History of Glaxo* (London: Profile Books 2001) pp. 385–414.

products which were likely to have a global market, and to make it more difficult to launch minor derivative local products into the British market.[6]

The reimbursement system which was introduced after the creation of the National Health Service was also designed in a way that encouraged innovation. The Voluntary Price Regulation Scheme (VPRS), later renamed the Pharmaceutical Price Regulation Scheme (PPRS), was a non-statutory arrangement aimed at securing 'fair and reasonable' prices. A key feature of the scheme was that prices were negotiated, not on individual products, but on the profits of the firm as a whole. The allowed rate of return was negotiated directly and was set higher for firms which invested in research and manufacturing facilities in the UK than for those which used the UK simply as a market for drugs developed and manufactured elsewhere.

The pricing system has been described as pro-innovation but non-protectionist. Its effect was 'to encourage the world's strongest pharmaceutical companies to make significant direct investments in the UK, thereby exposing domestic firms to the full competitive force of the industry's best players'.[7] Among the non-British companies which invested in the UK were Pfizer from the US, which built a large research facility at Sandwich in Kent, and another American firm, G. D. Searle, whose laboratory in High Wycombe was to become a leading centre of research in molecular biology (and later an important source of talent for UK biotech firms).

After a critical review of the industry's pricing practices by the Sainsbury Committee in 1967 some changes were made to the scheme, but they did not undermine the principle that the price of drugs had to be set at a level which gave an incentive to manufacturers to spend money on research. As the committee pointed out, 'in the absence of the prospect of "abnormal" profits, private industry would have no special inducement to undertake research to which attached an abnormal risk of failure'.[8]

There were periodic complaints about excessive drug prices—an extreme case was the row between Roche and the government over the pricing of the tranquillizers, Valium and Librium, in 1973—but the regulatory system

---

[6] Lacy Glenn Thomas, 'Implicit industrial policy: the triumph of Britain and the failure of France in global pharmaceuticals', *Industrial and Corporate Change* 3, 2 (1994). Subsequent writers have criticized the British regulatory system for being too permissive and allowing some unsafe drugs to reach the market. John Abraham and Courtney Davis, 'Testing times: the emergence of the practolol disaster and its challenge to British drug regulation in the modern period', *Social History of Medicine* 19, 1 (2006) pp. 127–47. A more sweeping criticism of the pharmaceutical industry and the regulatory system is contained in Ben Goldacre, *Bad Pharma* (London: HarperCollins, 2012).

[7] Thomas, 'Implicit industrial policy'. See also Leigh Hancher, *Regulating for Competition: Government, Law, and the Pharmaceutical Industry in the United Kingdom and France* (Oxford: Oxford University Press, 1989) pp. 191–227.

[8] *Report of the Committee of Enquiry into the relationship of the pharmaceutical industry with the National Health Service* (Cmd 3410, HMSO, 1967) p. 41.

provided an important element of stability for British and foreign-owned pharmaceutical companies.

The other government contribution to the success of the industry was its support, channelled principally through the Medical Research Council, for health-related scientific research. As measured by the number of Nobel Prize winners and by the share of world citations in medicine and the life sciences, British scientists have performed better in this field than their counterparts in other European countries (Figure 8.1). The quality of medical science in British universities was a source of strength for the industry, and it was reinforced by the close links between academic scientists and clinical practice. Medical schools in the UK have been more closely associated with universities than is usually in the case in Continental Europe, and science has had a higher status within the medical profession.[9]

The institutional framework which took shape in pharmaceuticals after the war amounted to what Lacy Glenn Thomas has called an implicit industrial policy.[10] As important as the things which government did to help the

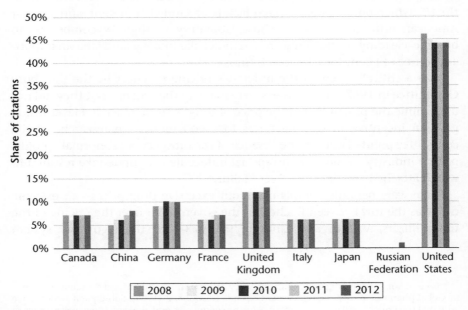

Figure 8.1. Share of life science academic citations

Source: Life Science Competitiveness Indicators, Department of Business, Innovation and Skills, March 2015 Chart 12B.

---

[9] Rebecca Henderson, Luigi Orsenigo, and Gary P. Pisano, 'The pharmaceutical industry and the revolution in molecular biology', in *Sources of Industrial Leadership*, edited by David C. Mowery and Richard R. Nelson (Cambridge: CUP, 1999) p. 279.

[10] Thomas, 'Implicit industrial policy'.

industry were the things that it did not do. It did not protect British-owned firms from foreign competition, as it did in computers; more than half of the drugs bought by the NHS came from non-British firms. It did not try to strengthen the industry by creating a national champion; when two of the leading firms, Glaxo and Beecham, proposed to get together, the transaction was banned by the competition authorities on the grounds that the industry needed multiple sources of innovation.[11]

A recent study attributed the post-war success of the British pharmaceutical industry to a combination of factors: outstanding science, much of it supported from public funds; a science-based tradition of healthcare; many brilliant individuals; shrewd business decisions; and a stable business environment.[12]

To this list should probably be added an element of luck. Glaxo would not have achieved its leadership position if it had not discovered Zantac; in the early 1990s this drug was still generating over 40 per cent of the company's sales. Drug discovery is an unpredictable process, with major breakthroughs being compared to 'black swan' events.[13] Very few of the drugs that enter clinical trials reach the market, and even fewer become blockbusters. Good fortune, as well as good science and good management, is part of the explanation for the success of individual firms in pharmaceuticals, and the same applies to biotechnology.

What was Special about Biotechnology?

A common justification for direct government intervention in a specific industry is the existence of market failures which, unless corrected by government, prevent technical or commercial opportunities from being fully exploited, with consequent damage to the economy. How to identify and remedy such failures is notoriously difficult. As two UK government economists have written, policy makers have only a fraction of the information available to market participants as a whole, so that any decision to intervene will be rough and ready. 'Not only is there a risk of miscalculating the effect of policy measures on their chosen target, or even choosing the wrong target altogether, but distortions may be created elsewhere in the economy.'[14]

---

[11] Monopolies Commission, *A report on proposed mergers between Beecham and Glaxo and between Boots and Glaxo* (HC341, HMSO, July 1972).

[12] Nicholas Owen, 'The pharmaceutical industry in the UK', in *Lessons from some of Britain's most successful sectors: an historical analysis of the role of government* (BIS Economics Paper No 6, March 2010).

[13] Bernard Munos, 'Lessons from 60 years of pharmaceutical innovation', *Nature Reviews Drug Discovery*, 8 (2009) pp. 959–68.

[14] John Barber and Geoff White, 'Current policy practice and problems from a UK perspective', in *Economic Policy and Technological Performance*, edited by Partha Dasgupta and Paul Stoneman (Cambridge: CUP, 1987).

Table 8.1. Market and institutional failures in biotechnology

- The time scales from initial scientific discovery to commercial exploitation are long
- The research involved is uncertain both in terms of outcome and who will benefit
- The research collaboration that is necessary between universities and firms is held back by coordination failure and a lack of familiarity and trust
- Because of organizational and strategic rigidities in the large pharmaceutical companies the development of new biotechnology-derived products in healthcare mainly rests on small entrepreneurial firms that lack business knowledge and skills
- Commercial uncertainties plus financial risks plus the small size of dedicated biotechnology firms inhibit the development of a steady stream of finance
- Lack of expertise in intellectual property management in research institutions inhibits the necessary transfer of technology to the commercial sector

Source: The Evaluation of DTI Support for Biotechnology, Main Report, Department for Business, Enterprise and Regulatory Reform, June 2008.

This comment was made in the mid-1980s when memories were fresh of some of the ill-judged interventions made by the Labour governments of the 1960s and 1970s. However, in the case of biotechnology, there has been a broad consensus among successive British governments, both Conservative and Labour, that because of market and institutional failures, and because of the importance of the sector, some government intervention is justified, to a degree that had not been necessary in pharmaceuticals.

A list of these failures was set out in an official report published in 2008, as part of an analysis of what government schemes to support biotechnology had achieved (Table 8.1). This report was written when a Labour government was in power and was more interventionist in tone than might have been acceptable to the Conservative governments of the 1980s and 1990s when the UK biotech sector was taking shape. Nevertheless, the report highlighted a question which all British governments have had to face: whether biotechnology was sufficiently special, and sufficiently important, to justify government intervention and, if so, what form that intervention should take.

Biotechnology Policy under Conservative Governments, 1979–1997

When Margaret Thatcher's Conservative government entered office in 1979 it was committed to reducing the role of the state in the economy; the apparatus of industrial intervention that had been created by the Labour government was soon dismantled. The use of public funds to create Celltech, as recommended by the Spinks committee, was a surprising deviation from the government's free market philosophy, justified on the grounds that it should help to promote closer interaction between academic science and industry. However, it was an isolated case, and there was no intention of repeating it. The public shareholding in Celltech was soon divested, and the company was not treated as a national champion in the way that ICL, the computer manufacturer, had been under the previous Labour governments.

Whether the formation of Celltech was a good way of kick starting the biotechnology sector has been a matter of dispute. Josh Lerner, a prominent US academic and policy adviser, regards it as an example of what governments should not do.[15] Special privileges and public money were given to a company whose management, Lerner argues, was 'manifestly incapable' of exploiting those resources. Despite building modern laboratories and hiring top-class scientists Celltech made little progress during its first 10 years in commercializing its technologies. Subsequent acquisitions of smaller British firms, according to Lerner, did nothing to strengthen the biotech sector.

It is true that Celltech was slow to bring drugs to the market, partly because it did not commit itself in its early years to a therapeutics-based business strategy. The government was probably wrong to have given Celltech exclusive rights to so many MRC discoveries; other firms might have exploited the technology more rapidly if they had been given the opportunity to do so. On the other hand, Celltech did to some extent blaze a trail which others followed. It recruited scientists and managers who moved on to senior positions in other firms, and some of the research programmes it started were in the end successful, albeit under different owners; Cimzia, Celltech's monoclonal antibody treatment for rheumatoid arthritis, passed into the hands of UCB in 2004, eventually reaching the market in 2008.

Celltech was a rare example of selective intervention at the firm level. The Thatcher government's assumption was that the emerging biotech sector would benefit from broader, non-selective policies aimed at encouraging entrepreneurship and innovation throughout the economy. The government's approach was to focus on diffusion-oriented or horizontal measures affecting all industries, rather than vertical measures that singled out particular industries for support.

This attitude was reflected in the government's lukewarm response to the other recommendations in the Spinks report.[16] The committee had called for greater coordination among the research councils and government departments which had an interest in biotechnology-related research. It proposed the creation of a director or coordinator who would have authority over the research councils and drive forward new initiatives in biotechnology.

---

[15] Josh Lerner, *Boulevard of Broken Dreams: Why Public Efforts to Boost Entrepreneurship and Venture Capital have Failed, and What to Do about It* (Princeton: Princeton University Press, 2009) pp. 145–56.

[16] The government's response was set out in a White Paper, *Biotechnology* (Cmnd 8177, HMSO, 1981).

This proposal was resisted by the Medical Research Council, which feared that resources would be diverted from its primary mission, which was to support basic research.[17] The government rejected the idea of a national programme for biotechnology research; each research council should make its own decisions about the appropriate balance between basic, strategic, and applied research. A biotechnology unit was formed within the Department of Trade and Industry (DTI), which kept in touch with biotech firms and looked for ways of helping them, but it had no authority over the research councils.[18]

Margaret Sharp, a science policy analyst, was critical both of the Spinks report, mainly for its failure to call for a big injection of funds into the biological sciences, and of the government's response.[19] In her view, far from providing strategic leadership as Spinks had recommended, the government had left the various agencies concerned with biotechnology to do their own thing. The danger was that the UK's efforts in biotechnology research would be spread too widely, and with too much duplication. She compared British biotechnology policy to 'a rudderless ship, launched but carried with the tide and lacking any sense of direction'.[20]

Ronald Coleman, head of the DTI's biotechnology unit, insisted that the government's policies were appropriate for biotechnology at its present stage of development. He pointed to the high level of investment which was being made in academic research and to the steps the government was taking to strengthen the links between universities and industry. All this, he said, increased the chances of British scientific research being converted into new medicines. Whether the opportunities were seized or not depended, not on the government, but on the effectiveness of UK companies in a free and competitive market. The government 'has a role in ensuring a favourable climate for enterprise, fair and even-handed regulations and an adequate science base providing sufficient skilled people to develop and commercialise

[17] For an account of the differing views within the research councils about biotechnology research see Brian Balmer and Margaret Sharp, 'The battle for biotechnology: scientific and technological paradigms and the management of biotechnology in Britain in the 1980s', *Research Policy* 22 (1993) pp. 463–78.

[18] The Department of Trade and Industry, formed in 1970, was the Whitehall department which handled the government's relations with industry. Broken up by Labour in 1974, it was re-established under the same name in 1983. The current incarnation of DTI is as the Department for Business, Innovation and Skills (BIS).

[19] Margaret Sharp, 'The management and coordination of biotechnology in the UK', *Philosophical Transactions of the Royal Society*, 324, 1224 (August 1989).

[20] Margaret Sharp, 'Biotechnology in Britain and France: the evolution of policy', in *Strategies for New Technology, Case Studies from Britain and France*, edited by Margaret Sharp and Peyer Holmes (Hemel Hempstead: Philip Allan, 1989) p. 156. See also Margaret Sharp, 'The management and coordination of biotechnology in the UK'.

the technology, but it strongly resists picking winners or identifying commercial goals'.[21]

The government recognized that if this market-based approach was to bear fruit, two problems had to be tackled. First, academic scientists had to move out of their ivory towers and collaborate more actively with commercial firms. Second, the supply of finance to entrepreneurs in science-based industries, not just biotechnology, had to be improved.

In 1984 the government abolished the near monopoly which the British Technology Group (formerly the NRDC) had enjoyed over the commercialization of publicly funded research.[22] Influenced in part by the NRDC's much-criticized decision not to patent the Milstein–Köhler discoveries in monoclonal antibodies, the change gave universities the right to exploit the intellectual property arising from their scientists' discoveries; the rationale was the same as for the Bayh–Dole Act in the US, which had been passed in 1980. However, most universities did not have the resources or specialist capabilities to take advantage of their new freedom, and it was not until the mid-1990s that US-style university spin-outs began to emerge.

During the 1980s the Science and Engineering Research Council set up a Biotechnology Directorate which ran 'clubs' to bring together industry and academics in selected fields such as protein engineering.[23] But sectoral initiatives of this sort did not play a large role in the government's technology policy. The LINK scheme, introduced in 1986, provided modest financial support for pre-commercial collaborative research between businesses and universities; the scheme was targeted mainly at small and medium-sized enterprises. Later renamed the Collaborative R&D Grants scheme, it provided support both for basic research and for near-market development; the firms that received these grants had to cover part of their costs from their own resources. A subsequent evaluation concluded that LINK improved the recipient firm's ability to manage innovation and enhanced its credibility in the eyes of potential investors.[24]

Margaret Thatcher, who had taken a degree in chemistry at Oxford and worked as an industrial chemist, believed that publicly funded science should be geared more directly to the needs of the economy, and that industrialists

---

[21] R. F. Coleman, head of the biotechnology unit in the Department of Trade and Industry, writing in 'Biotechnology policy and achievement, 1980–1988', *Philosophical Transactions of the Royal Society, Biological Sciences*, 324, 1224 (August 1989).

[22] The assignment to Celltech of commercial rights to the MRC's biotech-related inventions was a notable exception to NRDC's monopoly.

[23] Margaret Sharp, 'Biotechnology and advanced technological infrastructure policies: the example of the UK's protein engineering club', *Economics of Science, Technology and Innovation* 7 (1996) pp. 217–46.

[24] Khaleel Malik, Luke Georghiou, and Hugh Cameron, 'Behavioural additionality of the UK SMART and LINK schemes', in *Government R & D Funding and Company Behaviour* (Paris: OECD, 2006).

should be given more influence in deciding research priorities.[25] This approach was taken further in the early 1990s, after Mrs Thatcher had left office but while the Conservatives were still in power. In 1993 the government, as part of a review of science policy, set up a Council for Science and Technology which included representatives from industry as well as the universities, and was chaired by William Waldegrave, the minister responsible for science policy.[26] Although some scientists deplored what they saw as the utilitarian approach to academic research, it set science policy in a direction which continued without much change after the Labour government took office in 1997.

On access to finance, the government's first step, aimed at all new firms, was to introduce the Business Start-up scheme, later renamed the Business Expansion Scheme, which gave tax incentives to encourage private investors to put capital into unquoted firms, initially for a minimum period of 5 years. The scheme was only partially successful; the investment criteria were defined too broadly, enabling investors to use the scheme as a tax-efficient means of investing in property. It was replaced in 1993 by the Enterprise Investment Scheme, which proved to be an attractive vehicle for business angels, acting as investors and mentors in small entrepreneurial firms. The EIS was followed in 1995 by the introduction of Venture Capital Trusts, which enabled investors to invest indirectly in young firms through managed trusts. An evaluation of these schemes concluded that the EIS and VCT investments had a positive but modest impact on recipient firms.[27]

Under the SMART scheme (Small Firms Merit Award for Research and Technology), launched in 1986, firms with less than fifty employees could compete for government grants up to £45,000 to support technology projects; awards were made on the basis of technical and market feasibility. Although the amounts of money were small, the award of a SMART grant was seen as an accreditation of the firm's technology, and allowed it to advance to the point where private investors might take an interest.[28]

In line with the government's non-selective approach, the LINK and SMART schemes were not aimed at any particular industry. There was, however, some

[25] For a critical assessment of this approach, see David Edgerton and Kirsty Hughes, 'The poverty of science: a critical analysis of scientific and industrial policy under Mrs Thatcher', *Public Administration*, 67 (Winter 1989) pp. 419–33.

[26] UK Government, *Realising our Potential: a strategy for science, engineering and technology* (Cmd 2250, HMSO, 1993).

[27] Marc Cowling, Peter Bates, Nick Jagger, and Gordon Murray, 'Study of the impact of Enterprise Investment Scheme and Venture Capital Trusts on company performance', HM Revenue & Customs Research Report 44 (London, 2008). A later study showed that technology-focused VCT investment declined after the dot.com boom and bust and did not recover. Josh Siepel, 'Capabilities, policy and institutions in the emergence of venture capital in the UK and the US', D Phil thesis, University of Sussex, Brighton 2009.

[28] Bank of England, 'The financing of technology-based small firms' (October 1996).

support for enabling technologies which were expected to have wide impact throughout the economy. The largest programme of this sort was in information technology. Kenneth Baker, a Conservative MP who had argued strongly for a government initiative in this field, was appointed Minister for Information Technology in 1981. He played an important role in setting up the Alvey programme, which promoted collaboration between universities, industry, and government in advanced information technology research.[29]

Although there was no similar programme in biotechnology, the Conservative governments which held office between 1979 and 1997 recognized the importance of the new genetic engineering techniques, especially at the level of basic research. This was reflected in the reorganization of the research councils in 1993, which led to the creation of a new body, the Biotechnology and Biological Sciences Research Council (BBSRC). Support for medical research remained with the Medical Research Council, but the new Council had the main responsibility for funding basic research in biological sciences.[30] The first head of the BBSRC was Professor Sir Tom Blundell, who later co-founded Astex Technology (a biotech firm discussed in Chapter 4).

The exploitation of research was seen by government as a matter for the private sector. In biotech, however, the government was aware of the market and institutional failures that hampered the growth of new firms and took some steps to remedy them. These included the creation of the Biotechnology Finance Advisory Service, which helped small firms find sources of finance, and the Biotechnology Mentoring and Incubator Programme, which provided specialist business advice and laboratory space for start-up firms.[31] These schemes, administered by the Department of Trade and Industry, were extended by the Labour government after 1997.

Technology policy under Margaret Thatcher and her Conservative successors was criticized by some academic observers for putting too much stress on promoting entrepreneurial firms and not enough on creating 'durable technological capabilities'.[32] Any support for innovation had to meet strict value-for-money criteria, with the result, according to the critics, that

---

[29] The post of Minister of Information Technology was discontinued in 1987 and there was no follow-up to the Alvey programme, although some of its work continued in a new information engineering directorate within the DTI. Brian Oakley and Kenneth Owen, *Alvey, Britain's Strategic Computing Initiative* (Cambridge, MA: MIT Press 1989) pp. 245–64.

[30] William Lea, 'Reorganisation of the Science Research Councils', House of Commons Library, Science and Environment Section, Research Paper 94/19 (House of Commons, 1 February 1994).

[31] Details of these schemes are set out in Biopolis, *Inventory and analysis of national public policies that stimulate research in biotechnology, its exploitation and commercialisation by industry in Europe in the period 2002–2005, National Report of United Kingdom* (European Commission, March 2007), and in *The evaluation of DTI support for biotechnology, Main Report* (Department for Business, Enterprise and Regulatory Reform, June 2008).

[32] William Walker, 'National innovation systems: Britain', in *National Innovation Systems, a Comparative Analysis*, edited by Richard R. Nelson (New York: OUP, 1993).

long-term developments with uncertain payback were starved of money. However, as things stood in the mid-1990s, the Conservatives could claim that, thanks to their policies—broad support for entrepreneurship and innovation, together with a modest amount of targeted assistance for start-up firms—the biotech sector had got off to a promising start. New firms were being created in reasonable numbers; the venture capital industry was growing; and, at least until the implosion of British Biotech in 1997, investors had been showing a strong appetite for listed firms.

In an assessment of the pharmaceutical and biotech industries published in 1997 a *Financial Times* commentator noted that employment in pharmaceuticals had been declining, partly because of the cutbacks that had followed the merger between Glaxo and Wellcome in 1995.[33] Two of the smaller firms, Boots and Fisons, had sold their pharmaceutical businesses to foreign acquirers, while Beecham had merged with SmithKline Beckman in the US.[34] In 1993 ICI demerged its pharmaceutical business into a separately listed company, Zeneca, but neither this transaction nor the 1999 merger between Zeneca and Astra of Sweden was expected to have any effect on employment.

These mergers were part of a restructuring of the global pharmaceutical industry, and they led to some factory closures as the enlarged groups took steps to eliminate duplication; soon after the merger with Glaxo, Wellcome's Beckenham research site was closed with the loss of more than a thousand jobs. For the UK, however, according to the *Financial Times*, any contraction in pharmaceuticals arising from mergers and acquisitions was likely to be offset by growing employment in biotech.

With the benefit of hindsight, this judgement was over-optimistic, and the ability of biotech firms to create new jobs was exaggerated, but it reflected the confidence that prevailed at that time in the performance of the sector. Margaret Sharp had complained that in contrast to France there was no vision in the UK's approach to biotechnology. The British, she wrote, had adopted a 'pragmatic, incrementalist view' of genetic engineering. 'Where the pay-off is uncertain and long-term, it is for industry, not government, to decide where money is well spent.'[35] Yet by the end of the Conservatives' term of office the UK biotech sector was well ahead of its French counterpart in the number and size of its companies, and in the number of drugs under development.[36]

[33] Daniel Green, 'Britain: in remarkably good health', *Financial Times*, 24 April 1997. The Glaxo/Wellcome merger was prompted in part by Glaxo's over-dependence on Zantac and the need to broaden its portfolio of drugs under development.

[34] Smith Kline and French acquired Beckman, a diagnostics company, in 1982. The new company, SmithKline Beckman, merged with Beecham in 1989.

[35] Margaret Sharp, 'Biotechnology in Britain and France', p. 152.

[36] Ernst & Young, *European Biotech 1997*, p. 17.

Continuity under New Labour

Tony Blair's New Labour government, which took office in 1997, had no intention of returning to the old policy of picking winners. In industrial policy, as in most other aspects of economic management, the new government left the Thatcher reforms intact. The main change, driven by the Chancellor of the Exchequer, Gordon Brown, was that science and technology were given a higher priority. Funding for the research councils was increased, and tax incentives were introduced to encourage the private sector to spend more on research. Under the R&D tax credit scheme, introduced in 2000, small firms were allowed to deduct 150 per cent of their R&D expenditure from taxable income; the scheme was later extended to larger companies. This initiative was prompted by growing concern about the low level of business-financed R&D as a proportion of GDP in the UK, relative to other industrial countries.[37]

The Labour government was as concerned as its Conservative predecessor to encourage entrepreneurial activity in high-technology industries. An early move was the introduction of the Enterprise Management Incentive (EMI), whereby small unquoted firms could offer the equivalent of stock options to attract and retain key employees. As Gordon Brown explained when he introduced the scheme, its aim was to reward risk and stimulate new enterprise at the cutting edge of technology. 'I want to recruit, motivate and reward Britain's risk-takers—the innovators capable of creating wealth and jobs in the Britain of tomorrow.'[38]

When Peter Mandelson, one of the architects of Labour's election victory, was appointed Secretary of State for Trade and Industry in 1998, he was eager for Britain to move towards a US-style entrepreneurial culture. (He was confirmed in that view by an early visit to Silicon Valley.) Mandelson's first White Paper contained no hint of old-style Labour policies; the focus was on 'making markets work better'. The White Paper included a provision for a public–private venture capital fund to encourage entrepreneurship.[39]

A high priority for the new government was to encourage universities to put more emphasis on entrepreneurship and to accelerate the commercialization of new technologies. One element in this programme was a partnership between MIT and Cambridge University, aimed at learning from MIT's experience in creating new firms. Another was the creation of the University

---

[37] Stephen R. Bond and Irem Guceri, 'Trends in UK BERD after the introduction of R & D tax credits', Oxford University Centre for Business Taxation, Working Paper (January 2012).

[38] House of Commons Debates, 9 March 1999, cc 177–179.

[39] Peter Mandelson, *The Third Man* (London: HarperCollins, 2001) pp. 265–6. The government's approach was set out in a White Paper published in 1998: 'Our competitive future: building the knowledge-driven economy', Department of Trade and Industry, December 1998.

Challenge Fund, through which universities could compete for funds to support research projects that had commercial potential. By 2003 nineteen universities, including the leading research-intensive universities such as Oxford, Cambridge, Edinburgh, and Imperial College, had created seed funds and were actively promoting the creation of spin-out firms.

The hope was that British universities could emulate what MIT and other US universities had achieved in this area, but the focus on spin-outs had some perverse consequences. Inexperienced but ambitious university technology transfer offices vied with each other to boost the number of spin-outs; the result was to produce too many weak firms which were unable to attract the commercial funding that they needed to take their drug programmes forward.[40]

While there was no large-scale selective intervention in the early years of the Labour government, Ministers took a close interest in knowledge-based industries, and in biotech in particular. There was anxiety at the end of the 1990s about the efforts which several European countries, especially Germany, France, and the Netherlands, were making to catch up with the UK in this field.

Lord Sainsbury, the Science Minister, launched a review of biotech clusters, prompted in part by Germany's Bioregio competition, which was designed explicitly to support the growth of US-style clusters.[41] His report highlighted the problems caused by planning restrictions, especially in Oxford, Cambridge, and London, and urged the Regional Development Agencies to promote 'Urban Networks for Innovative Cluster Areas' in their regional strategies. However, the government did not follow Germany's example (described in chapter 6), by providing subsidies for start-up firms out of public funds.

The dialogue between the government and the biotech sector intensified after the dot.com boom-and-bust of 2000–01. The Bioscience Innovation and Growth Team (BIGT), which included government officials as well as industry executives and academics, was given the task of formulating a strategic approach to the future of the UK's bioscience industry.[42] In his foreword to the first BIGT report Tony Blair, Prime Minister, said the bioscience industry was a British success story and had an exciting future. But the tone of the report was gloomy. It focused on two main issues, the supply of finance to biotechnology firms and the need for better collaboration with the National

---

[40] A government review of business–university collaboration, published in 2003, pointed out that almost a third of the universities that had created spin-outs in 2002 did not bring in external equity for any of their new companies. This suggested that some of the public funding invested in recent years had not been sufficiently focused on quality. Lambert Review of business–industry collaboration, final report (HM Treasury, December 2003) pp. 60–2.

[41] *Biotechnology clusters*, Report of a team led by Lord Sainsbury, Minister for Science, August 1999.

[42] Similar teams were set up in other industries, including aerospace and the motor industry.

Health Service, especially in the area of clinical trials and in the approval procedures for novel drugs.

The government was sympathetic to the complaint that there was a lack of support for innovation within the NHS. This was a central theme in the government's health research strategy, Best Research for Best Health, which was published in 2006.[43] It led to the creation of the National Institute for Health Research (NIHR), which became the research arm of the NHS. This new body gave a greater emphasis to staff training, infrastructure, and systems, to make the NHS more receptive to research and innovation. The setting of targets for clinical trial recruitment and commencement times ensured that the NHS would be more responsive to the needs of industry. Crucially, NIHR replaced R&D allocations that had previously been tied to individual hospital trusts with a new framework which allocated funds to specific activities. With an annual budget in excess of £1bn,[44] NIHR came to play a prominent role in opening the NHS up for research; in less than a decade after its foundation, the number of NHS patients engaged in clinical trials had more than tripled.

A less welcome innovation, from the industry's point of view, was the creation of the National Institute for Clinical Effectiveness (NICE), to advise on whether new medicines offered sufficient value for money to be adopted throughout the NHS. Since its establishment in 1999 the industry's fear has been that NICE slows down or prevents the introduction of new drugs in the UK. Decisions made by NICE, together with moves to make the pricing criteria used in the PPRS less generous, have become a recurring source of conflict between the industry, patient groups, and the government.

The Return of Industrial Policy

In the early years of New Labour the phrase 'industrial strategy' was generally avoided, since it carried connotations of the failed policies of the 1960s and 1970s. But there was a gradual shift to the view that, in an era of globalization and intense competition from low-wage countries, the UK needed to concentrate resources on industries which had the best prospect of competing successfully in world markets. It was seen as a legitimate role for government to identify these industries and to look for ways of making them stronger.[45]

---

[43] Department of Health, 'Best Research for Best Health: Introducing a new national health research strategy' (London 2096).

[44] National Institute for Health Research, Annual Report 2013/2014 http://www.nihr.ac.uk/documents/about-NIHR/NIHR-Publications/NIHR-Annual-Reports/NIHR%20Annual%20Report%202013_2014.pdf.

[45] For a review of recent developments in UK industrial policy and proposals for reform, see Nicholas Crafts and Alan Hughes, 'Industrial policy for the medium and long term', Future of Manufacturing Project, Evidence Paper 37, Government Office for Science, October 2013. See also

In 2004 the innovation support schemes run by the Department of Trade and Industry, including LINK and SMART, were brought together in a new body, the Technology Strategy Board; it was given enhanced status as an independent non-departmental agency in 2007. The purpose of the TSB, Lord Sainsbury explained, was 'to help our leading sectors to maintain their position in the face of global competition'.[46]

When Peter Mandelson was brought back into the government in 2008 to head the industry department (a post which he had held in 1998–2000), he argued for what he called 'a new kind of smart, targeted intervention'.[47] This was not picking winners in the old sense, but 'a broader, strategic industrial activism'. The need for a targeted approach, focused on industries where the UK held, or could reasonably expect to establish, a competitive advantage, was set out in a White Paper, 'New industries, new jobs'.[48]

Pharmaceuticals and biotech fell into the favoured category. In 2009 the government set up the Office for Life Sciences, led by a new Science Minister, Paul Drayson (now Lord Drayson—he had been given a peerage in 2004).[49] Before joining the government in 2005 as minister for defence procurement, Drayson had co-founded and run PowderJect, one of the most successful of the first-generation biotech firms.

The creation of the Office came shortly after the second BIGT report, which expressed serious concern about the plight of the biotech sector.[50] It called for a widening of the costs that were reimbursable under the R&D tax credit scheme, and for changes in the EIS and VCT schemes, neither of which were accepted by the government. The government also rejected the suggestion that there should be an independent inquiry to assess NICE's long-term impact on the uptake of new medicines in the NHS.

Nevertheless, Lord Drayson recognized that the industry faced a financial crisis. There was also concern about moves by Big Pharma companies to cut back their research and development in the UK. In July 2009 the Office published the Life Sciences Blueprint, which among other things announced

David Connell and Jocelyn Probert, 'Exploding the myths of UK Innovation Policy', Centre for Business Research, Cambridge, January 2010.

[46] Lord Sainsbury of Turville, *The Race to the Top: a review of the government's science and innovation policies* (London: HM Treasury, October 2007).

[47] Mandelson, *The Third Man*, p. 456.

[48] *New Industry, New Jobs: Building Britain's Future* (Department for Business, Enterprise and Regulatory Reform April 2009).

[49] The government focus in the 2000s on 'Life Sciences' or 'Biosciences' rather than biotechnology reflected a commitment to support the medical devices and the established pharmaceutical industry as well as emerging biotechnologies which had been the priority in the 1980s and 1990s.

[50] *The review and refresh of Bioscience 2015*, a report to the government by the Bioscience Innovation and Growth Team (Department of Business, Innovation and Skills, 2009).

the launch of an 'Innovation Pass', whereby patients with the greatest need would have access to innovative drugs without going through a NICE appraisal.[51] On finance, the government said it was considering the introduction of a 'patent box', whereby income earned from intellectual property would be subject to a preferential rate of tax; although this would benefit all science-based industries, the patent box would be particularly valuable to drug discovery firms which relied heavily on patents.

All this activity came shortly before the general election of 2010, which led to the replacement of Labour by a Conservative–Liberal Democrat coalition. By this time, in the wake of the financial crisis, the need for an active industrial policy was widely accepted across the political spectrum. The Business Secretary, Vince Cable (a Liberal Democrat), made no apology for his commitment to 'a partnership with business that goes much further than narrowly addressing market failures'.[52]

Cable's Conservative colleague, David Willetts, Minister for Universities and Science, identified 'Eight Great Technologies' which merited support from the government. These included regenerative medicine and a broader category consisting of life sciences, genomics, and synthetic biology.[53] Government support, Willetts argued, should not be confined to the funding of scientific research.[54] Until recently, he said, governments had favoured horizontal measures rather than vertical ones focused on particular sectors, but he advocated a mixed approach. 'Strong science and flexible markets is a good combination of policies. But, like patriotism, it is not enough. It misses out crucial stuff in the middle—real decisions on backing key technologies on their way from the lab to the marketplace. It is the missing third pillar to any successful high-tech strategy.' In this context funding for the Technology Strategy Board continued to make a major contribution to applied research; the TSB distributed over £400m in 2014/2015 to R&D programmes with strong industry involvement.[55]

---

[51] Life Sciences Blueprint, a statement from the Office for Life Sciences, July 2009.

[52] Vince Cable, 'Challenges and opportunities for a knowledge-based UK economy', in *Mission-oriented Finance for Innovation*, edited by Mariana Mazzucato and Caetano C. R. Penna (London: Policy Network, 2015).

[53] David Willetts, *Eight Great Technologies* (London: Policy Exchange, 2013). The other six technologies were: the big data revolution and energy-efficient computing; satellites and commercial applications of space; robotics and autonomous systems; agri-science; advanced materials and nano-technology; and energy and its storage.

[54] Willetts was influenced by a group of economists who were pressing for more government intervention at the near-market end of the research and development process. These included Mariana Mazzucato, whose book, *The Entrepreneurial State* (London: Anthem Press, 2013), emphasized the role played by the Federal government in the US in supporting innovation (a theme that is explored in Chapter 2).

[55] The Technology Strategy Board was renamed Innovate UK in 2014.

Willetts's approach to innovation policy was reflected in the creation in 2012 of the £180m Biomedical Catalyst Fund, jointly run by the Medical Research Council and the Technology Strategy Board. Its purpose was to support promising R&D projects, mainly in small biotech and medical technology firms. Funding was awarded either on a co-investment basis, with public and private money supporting the same project, or using government money to bring products closer to the point where the firm could attract funds from commercial sources. The biotech sector also benefited from the establishment of the TSB-funded Catapult Centres, which were designed to promote collaboration between scientists and industry in high-technology industries.[56] One of the first to be established was the Cell Therapy Catapult, based at Guy's Hospital in London.

These interventions formed part of a wide-ranging strategy for the life sciences, prompted partly by the need to strengthen the biotech sector but also by growing concern over the future of the pharmaceutical industry. Pfizer's decision in 2011 to close its Sandwich research laboratories, with the loss of 2,400 jobs, underlined the fragility of an industry which had long been regarded as one of the country's few world leaders. At the end of that year David Cameron, the Prime Minister, set out a number of measures aimed at making the UK 'a world-leading place for life sciences investment'.[57] These included the introduction of the patent box, which had been proposed by the Labour government. Although the economic case for the patent box was questioned by economists,[58] it was welcomed by the industry. When GlaxoSmithKline subsequently announced a £200m investment in advanced manufacturing facilities, the company claimed that the patent box had been a crucial factor in its decision.[59]

The government published a separate report focused on enhancing the role of the National Health Service as a partner with industry in promoting innovation.[60] To address delays in securing approval for new drugs, the establishment of an 'early access to medicines' scheme was proposed, through which in certain circumstances the NHS would be allowed to take up innovative drugs before they had obtained regulatory approval for market launch. Some drugs would be given a 'promising innovation

---

[56] The Catapult Centre programme stemmed from recommendations made by Hermann Hauser, a leading IT entrepreneur and venture capitalist, in a report to the Labour government shortly before the 2010 election (Hermann Hauser, *The Current and Future Role of Technology and Innovation Centres in the UK* (Department for Business, Innovation and Skills, 2010)).

[57] *Strategy for UK Life Sciences* (Department for Business, Innovation and Skills, December 2011).

[58] Rachel Griffith and Helen Miller, 'Consultation on the UK patent box proposal: a response' (Institute of Fiscal Studies, 2011).

[59] *Financial Times*, 22 March 2012.

[60] *Innovation Health and Wealth: Accelerating Adoption and Diffusion in the NHS* (Department of Health, December 2011).

medicine' designation, similar to the 'breakthrough therapy' category that had been introduced by the FDA in the US.[61]

Other measures aimed at making the NHS more innovation-friendly included the creation of fifteen academic health science networks to bring together universities, hospitals, and industry with the aim of facilitating collaboration, guiding development of products, or reshaping development pathways. Clinical trial information systems and the NHS Innovation score card were devised to provide metrics of performance to monitor progress in research and adoption.

The importance which the government attached to the life sciences was underlined in 2014 with the appointment of George Freeman, a Conservative MP who had previously been a venture capitalist and biotech entrepreneur, to the new post of Minister for Life Sciences, reporting both to the Department of Health and to the Department for Business, Innovation and Skills. One of Freeman's first acts was to set up an Innovative Medicines and Medtech Review, whose remit was to make recommendations to the government on 'how to speed up access for NHS patients to cost-effective diagnostics, medicines and devices'.[62]

Not directly targeted at biotech, but of great interest to founders of biotech firms and their investors, were changes in the Enterprise Investment Scheme. In 2012 George Osborne, the Chancellor, announced that investors backing companies through the EIS could reclaim 30 per cent of their investment from the tax bill, up from 20 per cent previously.[63] He also increased the size of company that could make use of the scheme, up to 250 employees. In addition, he introduced a 50 per cent rate of tax relief for investors backing very early-stage firms through a new Seed Enterprise Investment Scheme. Further changes, announced shortly before the election in May 2015, included a provision that, for knowledge-intensive firms, the employee limit for eligibility under the EIS and the Venture Capital Trust scheme would rise from 250 to 500.

With the Conservative victory in the 2015 election, it seemed likely that government policy would continue broadly unchanged. How far this would involve an expansion of the Biomedical Catalyst fund and other forms of direct support was uncertain. The new Business Secretary, Sajid Javid, was less enthusiastic about the phrase 'industrial strategy' than Vince Cable, his Liberal Democrat predecessor, since it gave the impression that the government

---

[61] Another government move was the launch in 2010 of a national Cancer Drugs Fund, which was to spend £200m–300m annually on drugs not reimbursable on the basis of NICE cost–benefit assessments. The fund, welcomed by industry, is less popular with health economists and some clinicians who suggest it discriminates against patients without cancer, and wastes NHS resources. Andrew Ward, 'NHS to stop funding 16 cancer treatments', *Financial Times*, 12 January 2015.

[62] Written House of Commons statement by George Freeman, 20 November 2014.

[63] *Financial Times*, 25 October 2012.

cared strongly about some sectors of industry and not about others.[64] As a 'small-state Thatcherite', Javid seemed unlikely to favour extensive government intervention in industry. It was not clear whether this change of tone would affect government policy towards the life sciences industry.

The Government's Changing Contribution to Biotech

Since the publication of the Spinks report in 1980, UK government policy towards biotech has been a mixture of horizontal measures aimed at innovative activity of all kinds and industry-specific or vertical measures in which biotech, either on its own or as part of the life sciences industry, has been targeted for special assistance from public funds. The former were favoured in the 1980s and 1990s (the creation of Celltech being a notable exception), while the latter become more important in the early 2000s, and even more so after the financial crisis of 2008–09.

In the first category, three consistent strands of policy, under Labour and Conservative administrations, and under the Coalition government of 2010–15, have been: support for the science base; the promotion of closer links between academic science and industry; and the use of tax incentives to encourage firms to spend more money on research and to induce investors to invest in early-stage firms. Although these are not sector-specific policies, they have had a special importance for biotech because of the sector's close links with academic science and the dependence of early-stage biotech firms on an adequate supply of risk capital.

As Figure 8.2 shows, UK government spending on health-related research is dwarfed by the funds made available by the US government to the National Institutes of Health. NIH funding, as discussed in Chapter 2, has been a major contributor to the success of the US biotech sector, and of the life sciences industry as a whole.[65] However, the amount spent by the UK government on health-related research in recent years has been higher than in France, Germany, and Japan. Since 2010, despite the pressure on public finances resulting from the financial crisis and the subsequent recession, the UK government has largely protected the science budget from the cuts that have applied to other areas.[66]

---

[64] Interview with Sajid Javid, *Financial Times*, 17 September 2015.

[65] The high priority which US governments have attached to biomedical research was reflected in the American Recovery and Reinvestment Act of 2009, which, as part of a wide-ranging programme aimed at lifting the US out of recession, allocated $10.2bn to the NIH to be spent in 2009 and 2010, in addition to the annual budget of close to $30bn.

[66] Graeme Reid, *Why should the taxpayer fund science and research?* (National Centre for Universities and Business, 2014).

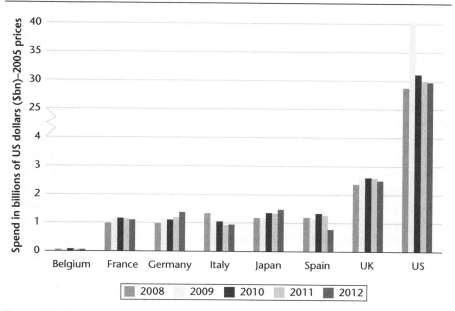

Figure 8.2. Government spend on health research and development

Source: Life Science Competitiveness Indicators, Department of Business, Innovation and Skills, March 2015, Chart 9.

Government finance has been supplemented by sizeable contributions from the Wellcome Trust and other non-government funders. One of the biggest recent investments has been the establishment of the Francis Crick Institute in London. Due to be fully operational in 2016, it will be one of Europe's largest biomedical research centres; it is backed by the Medical Research Council, Cancer Research UK, and the Wellcome Trust, together with Imperial College, University College London, and King's College London. Scientists in the Institute are expected to work closely with industry scientists in steering drug candidates from early-stage research into the clinic.[67]

As for the transfer of academic discoveries into industry, this was facilitated, first, by the Thatcher government's decision to end the monopoly which the NRDC had enjoyed over the commercialization of publicly funded research, and then by the resources provided by the Labour government after 1997 to enable universities to establish funds to support the creation of spin-out firms. Since then private-sector funds such as Imperial Innovations and the IP Group, as well as charities such as Cancer Research UK and The Wellcome

---

[67] One of the first such partnerships, with GSK, was announced in 2015. Each partner will contribute laboratory space and teams of around twenty scientists to work on early-stage projects. *Financial Times*, 14 July 2015.

Trust, have been increasingly active in identifying academic discoveries that have commercial potential, and in providing seed funding.[68] The result may have been an over-correction: there had been too few spin-outs before, but within a decade there were too many subscale firms being created and spun-out on their own too early.[69]

Biotech executives continue to complain that too many technology transfer offices are poorly managed. There is also a view that some universities see technology transfer as a means of generating extra funds, rather than as a way of exploiting academic innovations for the benefit of the wider economy.[70] Another problem, as seen by industry, is that the principal measure of academic success continues to be paper citations rather than the filing of patents. According to a survey carried out for the Wellcome Trust in 2014, UK academics are about half as likely to patent as their counterparts in a top US cluster, with nearly a third of UK respondents saying that their decision was based on the need for publications 'to drive grants or my career'.[71] Compared to the leading US centres, technology transfer in the UK still looks weak, but the responsibility for improving it now lies with the universities rather than the government.

On taxation, the EIS/SEIS schemes have been a useful device for encouraging high net worth individuals to invest in unlisted biotech firms. Several of the firms which have come to the fore in the last few years, such as Adaptimmune and Immunocore, have been financed wholly or mainly by private investors rather than venture capital. The value of the EIS lies in its stability—the scheme was introduced in 1993 and has survived several changes of government—and in the fact that, unlike government grant schemes, it is non-discriminatory; the qualifying criteria are relatively easy to understand and administer.

The industry continues to press for improvements in the EIS. For example, the annual limit of £5m for the amount a firm can raise in any one year through the EIS is thought to be too low for growing biotech firms. But the

---

[68] Another company that specialized in commercializing academic discoveries, Fusion IP, was acquired by IP Group in 2014.

[69] Differences in the views of studies undertaken 10 years apart are revealing. Commentators in the UK were discussing the proliferation of small, under-funded biotech firms (in the 2000s), whereas a decade earlier it was the weaknesses (in the 1990s) in the number of start-ups in areas like gene therapy that was a source of concern. See: Pari Patel, Antony Arundel, and Michael M. Hopkins, *Sectoral Innovation Systems in Europe: monitoring, analyzing trends, and identifying challenges in Biotechnology* (final report) (Europe Innova. European Commission, 2008); Paul Martin and Sandy Thomas, 'The "commercialization gap" in gene therapy: lessons for European Competitiveness', in *Biotechnology and Competitive Advantage: Europe's Firms and the US Challenge*, edited by Jacqueline Senker, coordinated by Roland van Vliet (Cheltenham: Edward Elgar, 1998) pp. 6–18.

[70] BioIndustry Association, *A vision for the UK life sciences sector in 2025* (London: April 2015).

[71] The UK's innovation ecosystem: summary of a review commissioned by the Wellcome Trust (November 2014).

scope for change is limited: the EIS has to strike an appropriate balance between providing incentives for investors and ensuring that the scheme is not abused for tax avoidance purposes.

Some biotech firms have suggested the introduction of a tax incentive for institutional investors—similar in principle to the EIS regime for private investors—which would encourage them to allocate part of their portfolio to early-stage firms. As noted in Chapter 6, how to mobilize pension fund money for investment in biotech is a continuing concern in other European countries as well as the UK. If any new incentive were to be introduced in the UK it could involve changes in the corporation tax arrangements as they affect pension funds, but the government has not so far shown any inclination to take up this suggestion.

More controversial is the question of how far private-sector funding should be supplemented by direct government support for individual firms. In the 1980s and 1990s, apart from the Celltech case, direct support was largely confined to small R&D grants, some of which were linked to collaborative projects with universities. Several of the regional development agencies, which were established in 1998 and abolished in 2010, backed local biotech firms in the hope of creating biotech clusters, generally working with local universities. These schemes may have had the effect of creating too many weak firms. As William Bains has remarked, 'They are formed not because there is a potent commercial reason for forming them but because you can get a grant for a new company, a university technology transfer office gets a tick in their appraisal form, or for some other reason. These companies are then kept afloat with more government money.'[72]

The recently established Biomedical Catalyst fund largely avoids these problems because it is focused on peer-reviewed applications to fund specific projects rather than companies, and involves co-investment with the grant recipient. The fund has been welcomed by many in the biotech sector; it has allowed firms to accelerate their research programmes and to broaden their pipelines.

As commonly happens with government schemes of this sort, there are complaints about bureaucracy and inconsistent judgements. There is also the danger that firms may put too much effort into winning grants rather than concentrating on commercial goals, and that government grants will keep the 'walking dead' alive.

As one experienced biotech investor pointed out, grant givers and grant recipients have different objectives and motivations. 'Grant givers like to see their awards as being additive, funding R & D which would not otherwise take

---

[72] William Bains, personal communication. For a critique of the promotion of new firms in biotechnology, see William Bains, *Venture Capital and the European Biotechnology Industry* (London: Palgrave Macmillan 2009) pp. 173–82.

place. Grant recipients want to get grants for work they are already undertaking or plan to undertake, thereby gaining additional resources for the firm. So the application and award process becomes something of a dance with each side obfuscating, trying to please the other. Since all awards involve co-investment by the recipient the best kind of awards to win are the ones where the recipient has already done the work.'[73]

Given the constraints imposed by European Union rules on state aid to industry, the scope for a substantial expansion of the Biomedical Catalyst fund is limited. For some biotech executives, the most useful contribution to the success of UK biotech lies, not in the provision of finance, but in making the NHS a more receptive customer for innovative medicines. Andy Richards, one of the country's most successful biotech entrepreneurs, put the point bluntly in evidence to a House of Commons Committee. 'It is notoriously hard to sell anything new into the NHS. This is partly a cultural thing. Partly, there are some elements within the NHS that for one reason or another have a "they shall not pass" mentality. It does make it incredibly hard to innovate in the medical field in an environment where your local market is hard to access.'[74]

The 'early access to medicines' scheme discussed earlier was seen as a step towards removing this obstacle, and the review set up by George Freeman may lead to other changes in the same direction. There is no doubt that because of the rigorous cost–benefit appraisals by NICE, changes in the Pharmaceutical Price Regulation Scheme, and the financial pressures on the NHS, the UK has become a less attractive market in recent years, especially for expensive biological drugs; such drugs are more regularly prescribed in other European countries such as France and Germany than in the UK. One consequence is that doctors in the UK are less well informed about novel drugs and hence less useful to local biotech firms.

On the other hand, the UK represents only about 3 per cent of the world demand for medicines (probably less for biologics). When biotech firms are trying to decide which R&D projects to pursue, the UK market does not generally figure prominently in their thinking, and the same applies to investors. Making the NHS more innovation-friendly would be welcomed by biotech firms, and the NHS could be more helpful to the life sciences industry not only as a customer but also as a partner—for example, as a source of data and in the conduct of clinical trials. However, for ambitious drug discovery firms that want to be world players, the principal focus has to be on international markets and above all the US.

---

[73] Private communication.
[74] Andy Richards, Evidence to House of Commons Science and Technology Committee, *Bridging the Valley of Death: Improving the commercialisation of research*, 18 April 2012.

The Limits of Industrial Policy

Over the past decade the concept of industrial policy has staged something of a comeback, in Europe and in other industrial countries. In the UK this is partly a response to the financial crisis, which highlighted the risks of the country's over-dependence on financial services and the need to shift resources into knowledge-based industries—whether in manufacturing or services—which have the potential to grow and are less vulnerable than older industries to competition from newly industrializing countries. Several economists have urged governments, and the European Commission, to look more favourably on sector-specific initiatives, especially in technologies that have a wide economic impact.[75] These economists recognize that errors were made by several countries in the conduct of industrial policy in the early post-war decades; there was too much emphasis on picking winners, many of which turned out to be losers, and too much concern with the creation of national champions. But they believe that carefully designed industrial policies, giving more weight to competition, can contribute to the necessary restructuring of European economies.

In their approach to biotech, successive British governments have in general steered clear of repeating the mistakes made in the 1960s and 1970s. One may question the wisdom of creating Celltech, but it was not a national champion in the old sense, and it did not seriously distort competition in the industry. Governments did not protect British biotech firms from foreign competition or erect barriers against foreign takeovers. To that extent there were some similarities with the policies pursued towards the pharmaceutical industry in the earlier post-war decades. But the international environment for biotechnology firms was very different from what it had been for the British pharmaceutical companies.

Although there was strong international competition in pharmaceuticals in the 1960s and 1970s, and a good deal of cross-border investment, national markets were segmented by different regulatory arrangements and different pricing regimes. It was possible for national governments to use regulatory policy to influence investment decisions, both by their own companies and by foreign pharmaceutical manufacturers. This is what the British government did, as discussed earlier in this chapter, partly through the pricing regime. It was an implicit rather than an explicit industrial policy, and there were many other factors, including the quality of academic science, which made the UK an attractive place for doing research in the life sciences. During those years it

---

[75] Philippe Aghion, Julian Boulanger, and Elie Cohen, 'Rethinking industrial policy' (*Bruegel Policy Brief*, June 2011). See also Mariana Mazzucato, *The Entrepreneurial State*; Crafts and Hughes, 'Industrial policy for the medium to long-term'.

was possible for a pharmaceutical company to launch a drug successfully in its local or regional market and then to do international licensing deals.

By the 1990s and early 2000s the character of the world pharmaceutical market had changed. Competition became much more global, with a strong orientation towards the US. The costs of drug discovery rose sharply, and firms based in a medium-sized country like the UK could no longer hope to recoup their research costs from the domestic market alone. Another factor has been the dominant influence on drug approval of the two main regulatory bodies, the Food and Drug Administration in the US and the European Medicines Agency; the latter has played an increasingly important role, particularly for biologics regulation, in Europe over recent years.

Membership of the World Trade Organization and the European Union binds the UK to follow rules that limit the government's ability to support favoured domestic sectors. For example, in 2014 the UK government was forced to modify its much heralded patent box scheme in response to complaints from Germany and others that it led to artificial shifting of investment between European countries.

These international rules still leave governments with some room for manoeuvre in industrial policy. Many economists and scientists have argued that the UK should spend a higher proportion of its GDP on scientific research and training to support innovation.[76] Others suggest that government should make better use of its powers as a purchaser to support domestic industries, although cautious civil servants, focused more on value for money than on supporting small firms, have been reluctant to use procurement policy aggressively.[77]

In biotech, British governments have moved some way towards a more interventionist approach, and it will take some time before the full impact of these measures can be assessed; there is no clear link between recent government initiatives and the improvement in the fortunes of the UK biotech sector in 2014 and 2015, which followed the surge in the US market for biotech shares. Governments can and should continue to support scientific research and to facilitate its commercialization, but, given the global character of the industry, there is a limit to how far they can tilt the playing field in favour of their national firms through industrial policy, whether implicit or explicit.

---

[76] Reid, *Why should the taxpayer fund science and research?*; Tera Allas, *Insights from international benchmarking of the UK science and innovation system*, BIS Analysis paper No 03 (Department of Business, Innovation and Skills, January 2014).

[77] *Public procurement as a tool to stimulate innovation*, House of Lords Science and Technology Committee: First Report (House of Lords, 17 May 2011).

9

The Persistence of Industrial Leadership

> *...whether or not a national industry enjoys competitive advantages depends significantly on the extent to which the actions of governments, firms and other stakeholders are mutually reinforcing in creating an environment in which a particular industry can flourish.*[1]

This book has sought to explain how firms based in the US established a commanding position in biotechnology and to show why their rivals in other countries found it so difficult, despite substantial support in some cases from their governments, to match the success of the American leaders. The main focus has been on the UK, where the biotech sector started promisingly in the 1980s and 1990s but fell back after the turn of the century.

Compared to other new technologies, biotechnology is exceptional in the extent to which the first-mover nation, having created a new industrial sector in the 1970s and 1980s, widened its lead over other countries in the subsequent two decades and has continued to do so. The persistence of US leadership, as this book has shown, stems from a uniquely American innovation ecosystem, only parts of which could be transplanted to other countries.

A more common pattern in the evolution of industries is for countries which establish an early lead to be subsequently caught up by firms based in other countries, which learn from and sometimes improve upon the methods used by the first-movers. A classic example is the automotive industry. Henry Ford in the US invented mass production just before the First World War, and that technology was ideally suited to a large, homogeneous, and fast-growing domestic market. During the inter-war period and for some years after the Second World War General Motors and Ford were the world's largest, most productive, and most profitable car manufacturers.

---

[1] Johann Peter Murmann and Ralph Landau, 'On the making of competitive advantage: the development of the chemical industries in Britain and Germany since 1850', in *Chemicals and Long-term Economic Growth*, edited by Ashish Arora, Ralph Landau, and Nathan Rosenberg (New York: John Wiley, 1998).

By the 1960s firms in other countries were catching up. Japanese manufacturers, led by Toyota, adapted Fordist methods to the different conditions of the Japanese market—more fragmented than the US, with shorter production runs and a greater variety of models—and developed a manufacturing system that combined high productivity, reliability, and low cost.[2] This was the basis for the spectacular increase in Japanese car exports during the 1970s and 1980s. General Motors and Ford had to send teams of engineers to Japan to learn how to make cars the Japanese way.

In the automotive industry US leadership in the early years stemmed in part from the size and character of the domestic market. In other cases the source of competitive advantage, and often a more durable one, lies in national institutions, which provide firms with the skills and capabilities that they need to compete in their chosen field.

The rise of the German chemical industry in the late nineteenth century was based on a new technology—synthetic organic chemistry—which was applied first to dyes and then to other products including pharmaceuticals.[3] The entrepreneurs who exploited the technology drew on universities which, to a far greater extent than their counterparts in other countries, emphasized original scientific research as a central part of the professor's job. Firms such as Bayer, Hoechst, and BASF hired large numbers of scientists trained at well-funded German universities and used them to build up their own research laboratories.

This marked the start of the modern, science-based chemical industry, and German companies were the leading players up to and beyond the First World War. Towards the end of this period the US, the UK, and other countries began to reorganize their universities along German lines, with a greater emphasis on research. Several science-based companies emerged, such as DuPont in the US and Imperial Chemical Industries in Britain, and by the 1960s they ranked among the world leaders in the industry. The German chemical companies were still strong, but no longer dominant.

In biotechnology, by contrast, it has proved more difficult for firms outside the first-mover nation to match the success of the American pioneers. Part of the explanation lies in the circumstances surrounding the birth of the biotech sector in the US and in the events that followed in the rest of the world.

The US Innovation Ecosystem

The early US lead may have been partly a matter of luck—a scientific discovery made at a particular time and place, and the emergence of entrepreneurs with

---

[2] Michael A. Cusumano, *The Japanese Automobile Industry: Technology and Management at Nissan and Toyota* (Cambridge, MA: Harvard University Press, 1985).

[3] Johann Peter Murmann, *Knowledge and Competitive Advantage* (Cambridge: CUP, 2003).

the skill and vision to exploit it—but the reasons for the subsequent growth of the biotech sector lie in a set of institutions and policies which reinforced each other and allowed firms to exploit the knowledge and capabilities stemming from the US science base. Firms growing within this innovation ecosystem enjoyed a formidable competitive advantage, particularly given the scale of resources in the US devoted to biomedical research and healthcare.

When the first US biotechnology firms were formed in the 1970s some of the institutional support which they needed was already in place. The US had a large network of high-quality universities and hospitals whose scientific research, in molecular biology and other relevant disciplines, was generously funded by the Federal government; a venture capital industry which had experience of starting and nurturing new firms in other high-technology industries; and a stock exchange, NASDAQ, which had already established itself as an attractive home for young, entrepreneurial firms exploiting novel technologies. In addition, the US market for medicines was not only much bigger than that of other countries, but also, partly because of the absence of price controls, more rewarding for manufacturers of innovative drugs; this had helped to foster a large and profitable pharmaceutical industry, which was to become a source of capital and skills for biotech firms.

These existing assets were reinforced by policy changes which were helpful for the growth of the biotech sector. They included: the Bayh–Dole Act, which facilitated the transfer of technology from academia into industry; the Supreme Court ruling in the Chakrabarty case, which allowed genetically modified organisms to be patented; the establishment of a regulatory framework for the conduct of genetic engineering research; orphan drug legislation, which increased the rewards for firms that developed treatments for rare diseases; and a change in the rules governing pension funds, allowing them to invest on a large scale in venture capital.

This supportive institutional framework was in place by the early 1980s, just as investors were becoming aware of the progress which biotech firms were making in developing drugs. Although share prices fluctuated in the course of the decade, there were enough successes from the new firms to encourage investors to take an interest in the sector. Some of the successes were based on the use of recombinant DNA to produce known proteins such as insulin and human growth hormone, and these drugs quickly received regulatory approval. Genentech, for example, was founded in 1976, went public in 1980, and had its first drug approved by the FDA in 1982. Firms that came later, developing novel drugs with novel technologies, faced greater difficulties in bringing products to the market. There were many failures, not least among firms that were trying to commercialize monoclonal antibodies. But by the end of the 1990s several antibody-based drugs for cancer and other diseases had been approved.

By this time the biotech sector was acquiring scale and momentum, with a sufficient number of revenue-generating firms and lucrative trade sales to attract investors. The stream of firms coming out of universities and research institutes, and the willingness of venture capitalists and other investors to support them, made for intense competition in the industry and ensured that discoveries with commercial potential were quickly exploited by well-resourced firms. When firms were acquired or went out of business, their managers and scientists, part of a large, highly skilled, and flexible workforce, could quickly find employment in another start-up.

The world financial crisis that began in 2008 led to a temporary halt in IPOs, but US investor interest soon revived and by 2012 the groundswell of support had brought a surge in biotech flotations as well as boosting the shares of established firms. Investor enthusiasm almost certainly went too far, and the shares of some of the newly listed firms rose to what appeared to be unrealistic heights. But the boom, which continued into 2015, had a more solid basis than the so-called genomics bubble of 1999–2000. The publicly listed biotech firms were now profitable in aggregate, although most were still loss-making; the pace of FDA approvals was accelerating; and there was new excitement around technologies, such as immunotherapy for the treatment of cancer, that seemed likely to generate big-selling drugs.

What Held Other Countries Back?

The rise of biotech in the US prompted scientists and entrepreneurs in other countries to launch biotech ventures on the American model. These efforts were generally supported by governments, which believed that the new techniques (not only in medicine but also in other fields) would create new jobs, boost exports, and contribute to economic growth. Although the economic impact of biotech was not as immediate or as extensive as many had hoped, the sector has continued to figure prominently in government programmes aimed at encouraging the growth of science-based industries.[4]

Among European countries the UK was in some respects well qualified to follow the American lead. Although government support for scientific research was much smaller than in the US, the quality of academic science in molecular biology and genetics was on a par with the US. During the 1980s and 1990s some of the institutional support which the emerging biotech

---

[4] Michael M. Hopkins, Paul A. Martin, Paul Nightingale, Alison Kraft, and Surya Mahdi, 'The myth of the biotech revolution: an assessment of technological, clinical and organizational change', *Research Policy* 36 (2007) pp. 566–89.

sector needed, including venture capital and more permissive stock market rules for early-stage firms, was established in the UK.

The UK seemed on the way towards establishing an investment infrastructure comparable to that of the US, with growing support for biotech firms from banks, brokers, and analysts. But while investor support for biotech was achieved it was not sustained. Whether through naiveté or inexperience on the part of company founders and investors, too many firms were floated on the stock market too early, on the basis of unrealistic projections about how soon they would bring drugs to the market. Following a series of corporate setbacks in the late 1990s and early 2000s the sector lost momentum. It is only in the past few years, as discussed in Chapter 5, that the fortunes of the sector have begun to recover, after a lost decade.

Why did the UK not do better? The first generation US firms were pioneers, not only in scientific terms, but also in creating a new type of organization, one that integrated the skills of academic scientists and commercial managers in a way which, despite some failures, proved extraordinarily productive. That early start, and the successes that flowed from it, set in motion a virtuous cycle, whereby increases in investment, R&D activity, optimistic expectations, and new innovations fed off each other.[5] No such early successes were forthcoming in the UK. One could say that the UK suffered from making a false start, with ill-informed investors piling into the sector in the expectation of rewards that never came—except to a fortunate few who bought and sold their shares at the right time.

The UK was also perhaps unlucky in that its seminal contribution to the rise of biotechnology was the discovery of monoclonal antibodies rather than recombinant DNA. Recombinant DNA was invented in California. One of the inventors, Herbert Boyer, co-founded Genentech in San Francisco; it was a highly successful business and profitable investment. Other firms, on the East as well as the West Coast, followed the same path and they were able to draw on a large and rapidly growing pool of scientists trained in the new technology—a resource that existed to a much lesser degree in the UK. The first products coming out of US biotech firms have been described as low-hanging fruit, because their medical potential was less uncertain (although significant production challenges had to be overcome). Enough recombinant protein drugs were successful to attract more investors into the sector, and some firms from that first wave of start-ups, notably Amgen and Biogen, still rank among the leading firms in the industry.

---

[5] Such virtuous cycles are documented in other contexts—see M. P. Hekkert, R. A. A. Suurs, S. O. Negro, S. Kuhlmann, and R. E. H. M. Smits, 'Functions of innovation systems: A new approach for analysing technological change', *Technological Forecasting and Social Change* 74, 4 (2007) pp. 413–32.

US biotech firms have contributed to the development of many innovative biologics as well as small-molecule drugs. Over recent years, they have been producing more new classes of drugs than the established pharmaceutical firms.[6] This contrasts with the UK, where the development of new drugs is still dominated by Big Pharma and the output of biologics has been low. UK biotech firms sought to develop very few recombinant protein drugs. By the time the first British biotechnology firms came on the scene the low-hanging fruit had been picked or else US firms had established an unassailable lead.

In scientific terms, the UK was ahead of the US in monoclonal antibodies at the start, but it took much longer than expected to apply the technology safely and effectively in therapeutics. The first antibody-based drug, launched by Ortho, a subsidiary of Johnson & Johnson, in 1986, was an immunosuppressant used to reverse transplant rejection, but it was not until the mid-1990s that monoclonal antibody technology was exploited with significant commercial success, in treatments for diseases such as cancer and rheumatoid arthritis. Here again American firms, including Genentech, took the lead, drawing partly on technology derived from the UK, partly on knowhow developed in American universities. The first UK-originated monoclonal antibody drugs to be brought to market were Mylotarg (2000), Campath (2001), and Humira (2002); all were licensed and launched by US companies. As in the case of recombinant technology, few UK firms pursued the novel antibody techniques for drug development.

Instead of exploiting the major breakthrough biotechnologies from which the leading US firms have profited, UK biotechs for the most part have focused on small-molecule drugs derived from the older synthetic organic chemistry tradition. Some sought to develop drugs against novel targets, but as the case of British Biotech showed, it was all too easy to underestimate the difficulty of developing safe and effective drugs to act on entirely new physiological insights. Most UK firms taking this route foundered. (Oxford GlycoSciences' success with the orphan drug Zavesca remains a rare exception, and much of the profit went to Actelion of Switzerland, which licensed the drug.) Many UK firms chose to focus on incremental forms of innovation such as drug reformulation and drug delivery technologies, aiming to make small-molecule drugs that had already been approved more effective. These were strategies that the City of London's generalist investors had more faith in. Few UK firms were able to explore the opportunities offered by new technologies such as gene therapy, RNAi, or stem cell therapies.

---

[6] In the US, biotech firms dominate the development of innovative biologics, while the development of innovative small molecule drugs is more evenly shared between Big Pharma and biotech—based on data provided by Robert Kneller, 'The importance of new companies for drug discovery: origins of a decade of new drugs', *Nature Reviews Drug Discovery* 9 (2010) pp. 867–82.

It is unfair to describe the performance of UK biotech as a failure; it remains the largest sector in Europe and has the most drugs in clinical trials. Among individual firms, Shire has grown from its British origins to become a leading speciality pharma company, with the bulk of its operations in the US. BTG, born as a technology licensing company out of a merger between two government agencies, has followed a similar strategy to Shire and has served its investors well. Nevertheless, there is a widespread perception that the innovative science coming out of British universities and research institutes could have been exploited more effectively by national firms. There are no British counterparts to US firms as successful as Amgen, Biogen, or Gilead.

The UK's inability to emulate the US, in terms of generating large firms with successful medicines in an emerging industry, has often been attributed to distinctively British institutional weaknesses rather than unique US advantages. Chief amongst these apparent weaknesses is finance. It has been argued that fund managers in the UK, because they are under pressure to achieve high returns from their investments in the short term, are unwilling to support firms whose development programmes will only pay off, if they pay off at all, over 10 years or more. Yet, as earlier chapters have shown, at least some of the newly created biotech firms were able to obtain significant investment from public investors through the stock markets. It was only after a string of failures in the late 1990s and early 2000s that investors withdrew their support. The withdrawal was due more to the disappointing returns from their earlier investments than to short-termism.

Another possible explanation for the dearth of world-leading, British-owned biotech firms is that old standby of business historians—an entrepreneurial or managerial failure. It is often said that for cultural reasons British scientists are less entrepreneurial and less interested in making money than their US counterparts. There are continuing complaints from biotech firms that academic incentives give too much priority to 'ivory tower' behaviour—securing citations in peer-reviewed journals and winning research grants—and not enough to making an impact on the real world.[7] But as far as biotech is concerned there has been a steady flow of British scientists who have founded or co-founded biotechnology firms. There are repeat company founders like Gregory Winter and Steve Jackson in Cambridge and Steve Davies in Oxford, and serial entrepreneurs like Andy Richards and Bryan Morton. The difference with the US is one of scale: there are many more such people on the other side of the Atlantic (including a good many British expatriates), and more of them have achieved outstanding success within the US innovation ecosystem.

---

[7] BioIndustry Association, *A vision for the UK life science sector in 2025* (London: April 2015).

As for the quality of management, some of the British firms that were founded in the 1980s and 1990s were run for too long by scientists with little commercial experience, or with too few colleagues around them who had the appropriate combination of managerial skills. But this is not unusual in the early phase of an industry which was attracting a good deal of speculative excitement and in which there was a high degree of inexperience and ignorance on the part of founders and investors. Looked at over a longer period, the sector has attracted some highly competent managers such as Peter Fellner and Ian Garland—the former a scientist by background, the latter trained as an accountant.

Could the government have done more to support British biotech firms? From the late 1970s onwards, as discussed in Chapter 8, successive British governments have taken a special interest in biotechnology and have supported the sector in various ways. The creation of Celltech in 1980 was an attempt to encourage the commercial exploitation of British scientific advances in molecular biology and genetics, and it was followed by a series of measures aimed at improving the environment for the creation and growth of firms in all high-technology industries.

Since the late 1990s government policy has become more interventionist, prompted in part by concern over the health of the British life sciences industry as a whole. A notable policy innovation was the creation in 2012 of the Biomedical Catalyst fund, designed to help firms navigate the so-called valley of death between early stage research and commercialization. Whether direct government support for near-market development should be taken further is a controversial issue, but most participants in the industry agree that any such support should not be at the expense of funding for basic scientific research in universities.

Governments have made some mistakes—for example, the over-enthusiasm for university-based spin-outs, which resulted in the creation of too many weak firms. But, taking the whole period from 1980 to 2015, it is hard to argue that policy errors on the part of government have seriously hampered the growth of the biotech sector.

The UK's experience in biotech over the past 30 years has been similar in some respects to that of other European countries. All of them faced the same challenge—how to create a framework of institutions and policies that would stimulate the growth of biotech firms. In some cases, like that of Germany discussed in Chapter 6, there were institutional obstacles, especially in access to finance, which had to be removed, and there is still concern in that country and others about the reluctance of large institutional investors to support the biotech sector.

A wider problem for European firms has been the absence of a genuinely integrated pharmaceutical market in Europe, comparable to that of the US.

Some steps towards greater integration have been taken with the creation of new European institutions—in the field of intellectual property rights, for example, and in regulation. Since the early 1980s there has also been considerable support, through the European Commission, for the funding of scientific research. But national governments continue to operate their own pricing and reimbursement arrangements, which makes the launch of new drugs slower and more cumbersome than in the US. For this and other reasons the US is a more attractive launching ground for innovative drugs than Europe.

If there were missed opportunities in UK biotech, they have to be seen in a wider European context. Other countries, including ones such as Germany and Switzerland which had scientific expertise and strong pharmaceutical industries, have struggled to foster a viable biotech sector.

The Future of the Biotech Sector

American capitalism is uniquely suited to the fostering of new firms in science-based or high-technology industries. In the case of biotech, key reinforcing factors have been the scale of government support for biomedical research, a strong emphasis on technology transfer between the research base and industry, access to national capital markets that are unrivalled in scale and specialist expertise, a large home market which is exceptionally receptive to novel drugs, and an established pharmaceutical industry, which already had a dominant global position before the rise of biotech.

European countries, even if they pooled their resources, cannot hope to match the US in its support for research and it is questionable whether they should do so. The huge amount of public money devoted to drug research and development in the US may well have diverted resources from other forms of health-related innovation. While life sciences is an industry in which European firms might be expected to have a competitive advantage, European governments need to guard against over-privileging drug development, in terms of access to public funds. In technology transfer, too, there is a risk that the pressure on universities to ensure that their research has an 'impact' will distort priorities and be detrimental to their primary function of advancing scientific knowledge.

The fragmented European market for drugs is considerably less attractive than the lucrative American drug market, which all biotech firms, particularly those based the US, benefit from. Yet the US approach to healthcare policy is far from perfect, and there are wasteful aspects of the system which should not be imitated in an effort to stimulate investment in European biotech. Nevertheless, the US model of entrepreneurial innovation is one from which other countries can learn, and have learned. As discussed in earlier chapters,

227

substantial reforms have taken place in many countries, including the UK. Institutional obstacles to the growth of biotech firms, such as the lack of capital for early-stage businesses and the absence of effective technology transfer arrangements, have been partially addressed. Progress has been hampered by the difficulty of transplanting institutions from country to country or even from sector to sector, especially where those institutions form part of an integrated system.[8]

Reforms in rival countries have not so far significantly reduced the magnetic attraction of the US—for biotech investors and pharmaceutical companies as well as for scientists and entrepreneurs. But past history suggests that the domination of a particular industry or set of technologies by firms based in a single country does not last indefinitely. The biotech sector as defined in this book is around 40 years old. New scientific and commercial opportunities are emerging at what appears to be an accelerating rate. Changes which the British and other European governments have made in the past few years suggest that some non-American firms are now better placed to take advantage of the new opportunities.

How many of them grow to become medium-sized or large companies, how long they stay independent, whether they have their IPOs in Europe or in the US—these are matters over which national governments have only limited influence. Governments have no magic wand with which to produce world-leading companies or world-beating products, but it is in their power to establish a stable business environment with appropriate supporting institutions. First and foremost is an adequately funded science base, generating the high quality research and essential skills on which the industry relies.

[8] Richard R. Nelson, 'What enables rapid economic progress: What are the needed institutions?' *Research Policy* 37 (2008) pp. 1–11.

Bibliography

Abraham, John, 'Pharmaceuticalization of society in context', *Sociology* 44, 4 (2010) pp. 603–22.

Abraham, John and Courtney Davis, 'Testing times: The emergence of the practolol disaster and its challenge to British drug regulation in the modern period', *Social History of Medicine* 19, 1 (2006) pp. 127–47.

Achilladelis, Basil, 'Innovation in the pharmaceutical industry', in *Pharmaceutical Innovation, Revolutionising Human Health*, edited by Ralph Landau, Basil Achilladelis, and Alexander Scriabine (Philadelphia: Chemical Heritage Press, 1999).

Ackerly, Dana T. 'Easdaq—the European market for the next hundred years?' *Journal of International Banking Law* 12, 3 (1987) pp. 86–91.

Adelberger, Karen E. 'A developmental German state? Explaining growth in German biotechnology and venture capital', BRIE Working Paper 134 (The Berkeley Round-table on International Economy, University of California, Berkeley, 1999).

Aghion, Philippe, Julian Boulanger, and Elie Cohen, *Rethinking industrial policy* (Bruegel Policy Brief, June 2011).

Allansdottir, Agnes, Andrew Bonaccorsi, Alfonso Gambardella, Myriam Mariani, Luigi Orsenigo, Fabio Pammolli, and Massimo Riccaboni, *Innovation and competitiveness in European biotechnology*, Enterprise Papers No 7 (Enterprise Directorate-General, European Commission, 2002).

Allas, Tera, *Insights from international benchmarking of the UK science and innovation system*, BIS Analysis Paper No 03 (Department for Business, Innovation and Skills January 2014).

Arcot, Sridhar, Julia Black, and Geoffrey Owen, *From local to global: the role of AIM as a stock market for growing companies* (London Stock Exchange, September 2007).

Armstrong, Sue, *P53: The Gene that Cracked the Cancer Code* (London: Bloomsbury Sigma, 2013).

Bains, William, 'What you give is what you get: investment in European biotechnology', *Journal of Commercial Biotechnology* 12, 4 (July 2006) pp. 274–83.

Bains, William, *Venture Capital and the European Biotechnology Industry* (Basingstoke: Palgrave Macmillan, 2009).

Bains, William, Stella Wooder, David Ricardo, and Munoz Guzman, 'Funding biotech start-ups in a post-VC world', *Journal of Commercial Biotechnology* 20, 1 (January 2014) pp. 10–27.

Balmer, Brian and Margaret Sharp, 'The battle for biotechnology: scientific and techno-logical paradigms and the management of biotechnology in Britain in the 1980s', *Research Policy* 22 (1993) pp. 463–78.

Bank of England, *The Financing of Technology-based Small Firms* (Bank of England, October 1996).

Barber, John and Geoff White, 'Current policy practice and problems from a UK perspective', in *Economic Policy and Technological Performance*, edited by Partha Dasgupta and Paul Stoneman (Cambridge: CUP, 1987).

Bernstein Research, *Glorious Middle Age: The 2012 Biotech Rally, why it might continue and how to participate* (New York: Bernstein Research, October 2012).

Bialojan, Siegfried and Julia Schüler, 'Commercial biotechnology in Germany: an overview', *Journal of Commercial Biotechnology* 18, 1 (September 2003) pp. 15–21.

Binder, Gordon and Philip Bashe, *Science Lessons, What the Business of Biotech Taught Me about Management* (Boston: Harvard Business Press, 2008).

BioIndustry Association, *A vision for the UK life science sector in 2025* (London: April 2015).

Biopolis, *Inventory and analysis of national public policies that stimulate biotechnology research, its exploitation and commercialisation by industry in Europe in the period 2002–2005, Final report* (European Commission, March 2007).

Bioscience Innovation and Growth Team, *Bioscience 2015, Improving national health, increasing national wealth*, a report to government by the Bioscience Innovation and Growth Team (Department of Trade and Industry, November 2003).

Bioscience Innovation and Growth Team, *The review and refresh of Bioscience 2015*, a report to government by the Bioscience Innovation and Growth Team (Department for Business, Enterprise and Regulatory Reform, January 2009).

Biotechnology, report of a joint working party of the Advisory Council for Applied Research and Development, Advisory Board for the Research Councils and the Royal Society (London: HMSO, 1980).

Booth, Bruce L., 'Beyond the biotech IPO: a brave new world', *Nature Biotechnology* 27, 8 (August 2009) pp. 705–9.

Branciard, Anne, 'France's search for institutional schemes to promote innovation: the case of genomics', Paper presented at the workshop on innovation, industry and institutions in France, (Paris: CEBREMAP, February 2000).

Breiding, R. James, *Swiss Made, the Untold Story behind Switzerland's Success* (London: Profile Books, 2012).

Bud, Robert, *The Uses of Life, a History of Biotechnology* (Cambridge: CUP, 1993).

Bud, Robert, *Penicillin, Triumph and Tragedy* (Oxford: OUP, 2007).

Bud, Robert, 'From applied microbiology to biotechnology: science, medicine and industrial renewal', *Notes & Records of the Royal Society*, 64 (July 2010) pp. 17–29.

Cable, Vince, 'Challenges and opportunities for a knowledge-based UK economy', in *Mission-oriented Finance for Innovation*, edited by Mariana Mazzucato and Caetano C. R. Penna (London: Policy Network, 2015).

Cambrosio, Alberto and Peter Keating, *Exquisite Specificity, the Monoclonal Antibody Revolution* (Oxford: OUP, 1995).

Casper, Steven, 'Institutional adaptiveness, technology policy and the diffusion of new business models: the case of German biotechnology', *Organisation Studies* 21, 5 (2000) pp. 887–914.

Casper, Steven, *Creating Silicon Valley in Europe, Public Policy towards new Technology Industries* (Oxford: OUP, 2007).

Casper, Steven, 'How do technology clusters emerge and become sustainable? Social network formation and inter-firm mobility within the San Diego biotechnology cluster', *Research Policy* 36 (2007) pp. 438–55.

Casper, Steven and Anastasios Karamanos, 'Commercialising science in Europe: the Cambridge biotechnology cluster', *European Planning Studies* 11, 7 (2003) pp. 805–22.

Cockburn, Iain M, Rebecca Henderson, Luigi Orsenigo, and Gary P. Pisano, 'Pharmaceuticals and biotechnology', in *US Industry in 2000, Studies in Competitive Performance*, edited by David Mowery (Washington DC: National Academy Press, 1999).

Cockburn, Iain M. and Scott Stern, 'Finding the endless frontier: lessons from the life sciences innovation system for technology policy', *Capitalism and Society*, 5, 1 (2010).

Cockburn, Iain M. Scott Stern, and Jack Zansher, 'Finding the endless frontier: lessons from the life sciences innovation system for energy R & D', in *Accelerating Energy Innovation: Lessons from Multiple Sectors*, edited by Rebecca Henderson and Richard G. Newell, National Bureau of Economic Research Conference Papers (Chicago: University of Chicago Press, 2011).

Cockburn, Iain M., 'The changing structure of the pharmaceutical industry', *Health Affairs* 23, 1 (2004).

Cohen, Ronald, *The Second Bounce of the Ball* (London: Weidenfeld and Nicolson, 2007).

Coleman, R. F., 'Biotechnology policy and achievement,1980–88', *Philosophical Transactions of the Royal Society, Biological Sciences*, 324 (31 August 1989).

Connell, David and Jocelyn Probert, *Exploding the Myths of UK Innovation Policy* (Centre for Business Research, Cambridge, January 2010).

Cooke, Philip, 'Biotechnology clusters in the UK: lessons from localisation in the commercialisation of science', *Small Business Economics* 17 (2001) pp. 43–59.

Coopey, Richard and Donald Clark, *3i, Fifty Years Investing in Industry* (Oxford: OUP, 1995).

Cortright, Joseph and Heike Mayer, *Signs of Life: The Growth of Biotechnology Clusters in the US* (Washington DC: Brookings Institution, June 2002).

Cowling, Marc, Peter Bates, Nick Jagger, and Gordon Murray, *Study of the impact of Enterprise Investment Scheme and Venture Capital Trusts on company performance* (HM Revenue & Customs Research Report No. 44, London, 2008).

Cox, George, 'Over-coming short-termism in British business: the key to sustained economic growth', an independent review commissioned by the Labour Party, February 2013.

Crafts, Nicholas and Alan Hughes, *Industrial policy for the medium to long-term*, Future of manufacturing project; evidence paper 37 (London: Office for Science, October 2013).

Cusumano, Michael A., *The Japanese Automobile Industry: Technology and Management at Nissan and Toyota* (Cambridge, MA: Harvard University Press, 1985).

Davies, Kevin, *The $1,000 Genome* (New York: Free Press, 2010).

Davis, Karen, Kristof Stremikis, David Squires, and Cathy Schoen, 'Mirror, mirror on the wall: how the US healthcare system compares internationally' (New York: The Commonwealth Fund, June 2014).

de Chadarevian, Soraya, *Designs for Life, Molecular Biology after World War II* (Cambridge: CUP, 2002).

de Chadarevian, Soraya, 'The making of an entrepreneurial science, biotechnology in Britain 1975–1995', *Isis* 202, 4 (December 2011) pp. 601–33.

Department for Business, Enterprise and Regulatory Reform, The evaluation of DTI support for biotechnology, Main Report, June 2008.

Department for Business, Innovation and Skills, *Strategy for UK life sciences* (December 2011).

Department of Health, *Innovation, Health and Wealth: Accelerating adoption and diffusion in the NHS* (December 2011).

Department of Trade and Industry, *Our competitive future: building the knowledge-driven economy* (December 1998).

Department of Trade and Industry, *The government response to the report by the Bioscience Innovation and Growth Team*, (London: HMSO, November 2004).

Department for Business, Enterprise and Regulatory Reform, *New industry, new jobs: building Britain's future* (April 2009).

Dickson, David, *The New Politics of Science* (Chicago: Chicago University Press, 1988).

Dodgson, Mark, *Celltech, the first ten years of a biotechnology company* (Science Policy Research Unit Discussion Paper, University of Sussex, 1990).

Dohse, Dirk and Tanja Staehler, 'BioRegio, BioProfile and the rise of the German biotech industry' (Kiel Working Papers No 1456, October 2008).

Dosi, Giovanni, Patrick Llerena, and Mauro Sylos Labini, 'The relationships between science, technologies and their industrial exploitation: an illustration through the myths and realities of the so-called European Paradox', *Research Policy* 35 (2006) pp. 1450–64.

Dwek, Raymond A., 'Journeys in science: glycobiology and other paths', *Annual Review of Biochemistry* 83 (2014) pp. 1–44.

Edgerton, David and Kirsty Hughes, 'The poverty of science: a critical analysis of scientific and industrial policy under Mrs Thatcher', *Public Administration* 67 (Winter 1989) pp. 419–33.

Ergas, Henry, 'The importance of technology policy', in *Economic Policy and Technological Performance*, edited by Partha Dasgupta and Paul Stoneman (Cambridge: CUP, 1987).

Ernst & Young, *Annual European Biotech Reports 1994–2003*.

Ernst & Young, *Annual Global Biotechnology Reports 2004–2015*.

Ernst & Young, *German Biotechnology Report 2011* (Ernst & Young, Mannheim, April 2011).

Ernst & Young and the UK BioIndustry Association, *Fundamental strengths of the UK ecosystem: state of the nation 2014* (London 2015).

Fisher, Lawrence M., 'The rocky road from start-up to big-time player: Biogen's triumph against the odds', *Strategy and Business* 8 (1 July 1987).

Fisken, Jane and Jan Rutherford, 'Business models and investment trends in the bio-technology industry in Europe', *Journal of Commercial Biotechnology* 8, 3 (Winter 2002) pp. 191–9.

Franzke, Stefanie A., Stefanie Grohs, and Christian Laux, 'Initial public offerings and venture capital in Germany', in *The German Financial System*, edited by Jan P. Krahmen and Reinhard H. Schmidt (Oxford: OUP, 2004).

Galambos, Louis and Jeffrey L. Sturchio, 'Pharmaceutical firms and the transition to biotechnology: a study in strategic innovation', *Business History Review* 72, 2 (Summer 1998) pp. 250–78.

Gambardella, Alfonso, *Science and Innovation, The US Pharmaceutical Industry during the 1980s* (Cambridge: CUP, 1995).

Giesecke, Susan, 'The contrasting roles of government in the development of biotechnology industry in the US and Germany', *Research Policy* 29 (2000) pp. 205–23.

Gittelman, Michelle, 'Mapping national knowledge networks: scientists, firms, and institutions in biotechnology in the United States and France', PhD Dissertation, University of Pennsylvania,(2000) p. 92.

Gittelman, Michelle, 'National institutions, public–private knowledge flows and innovation performance: a comparative study of the biotechnology industry in the US and France', *Research Policy* 35 (2006) pp. 1052–68.

Gold, E. Richard and Alain Gallochat, 'The European Biotech Directive: Past as prologue', *European Law Journal* 7, 3 (September 2001) pp. 331–66.

Goldacre, Ben, *Bad Pharma: How Drug Companies Mislead Doctors and Harm Patients* (London: Fourth Estate, 2012).

Gompers, Paul A., 'The rise and fall of venture capital', *Business and Economic History*, 23, 2 (Winter 1994).

Government Response to Bioscience 2015, Report by the Bioscience Innovation and Growth Team (Department of Trade and Industry, May 2004).

Government Response to Review and Refresh of Bioscience 2015 Report by the Bioscience Innovation and Growth Team (Department for Business, Enterprise and Regulatory Reform, May 2009).

Grabowski, Henry and John Vernon, 'Innovation and structural change in pharmaceuticals and biotechnology', *Industrial and Corporate Change* 3, 2 (1994).

Greene, Howard, Interview conducted by Matthew Shindell, 8 October 2008, The San Diego Technology Archive (SDTA), UC San Diego Library, La Jolla, CA.

Griffith, Rachel and Helen Miller, *Consultation on the UK patent box proposal: a response* (Institute of Fiscal Studies, 2011).

Hague, Douglas and Christine Holmes, *Oxford Entrepreneurs* (Oxford: Said Business School, 2006).

Hague, Douglas and Anthea Milnes, *From Bright Sparks to Brilliant Businesses, Oxford University Spin-offs and Start-ups* (London: Bloomsbury, 2011).

Hancher, Leigh, *Regulating for Competition: Government, Law, and the Pharmaceutical Industry in the United Kingdom and France* (Oxford: OUP, 1989).

Haslam, Colin, Nick Tsitsianis, and Pauline Gleadle, 'UK bio-pharma: innovation, re-invention and capital at risk', Institute of Chartered Accountants of Scotland (2011).

Hauser, Hermann, *The current and future role of technology and innovation centres in the UK* (Department for Business, Innovation and Skills 2010).

Hekkert, M. P., R. A. A. Suurs, S. O. Negro, S. Kuhlmann, and R. E. H. M. Smits, 'Functions of innovation systems: a new approach for analysing technological change', *Technological Forecasting and Social Change* 74, 4 (2007) pp. 413–32.

Henderson, Rebecca, Luigi Orsenigo, and Gary P. Pisano, 'The pharmaceutical industry and the revolution in molecular biology', in *Sources of Industrial Leadership*, edited by David Mowery and Richard R. Nelson (Cambridge: CUP, 1999).

Higgins, Monica C. *Career Imprints: How the 'Baxter Boys' Built the Biotech Industry* (San Francisco: Wiley, 2005).

Hopkins, Michael M., Philippa A. Crane, Paul Nightingale, and Charles Baden-Fuller, 'Buying big into biotech: scale, financing and the dynamics of UK biotech, 1980–2009', *Industrial and Corporate Change*, 22, 4 (August 2013) pp. 903–52.

Hopkins, Michael M., Paul A. Martin, Paul Nightingale, Alison Kraft, and Surya Mahdi, 'The myth of the biotech revolution: an assessment of technological, clinical and organizational change', *Research Policy* 36 (2007) pp. 586–9.

Hopkins, Michael M. and Joshua Siepel, 'Just how difficult can it be counting up R&D funding for emerging technologies (and is tech mining with proxy measures going to be any better?)', *Technology Analysis and Strategic Management* 25, 6 (2013) pp. 655–85.

House of Commons Science and Technology Committee, *Bridging the Valley of Death: improving the commercialisation of research*, Eighth Report of Session 2012–2013 (London: HMSO, March 2013).

Hughes, Alan, *Short-termism, impatient capital and finance for manufacturing innovation in the UK*, Future of manufacturing project: Evidence Paper 16 (Government Office for Science, London, December 2013).

Hughes, Sally Smith, 'Making dollars out of DNA; the first major patent in biotechnology and the commercialisation of molecular biology 1974–1980', *Isis* 92, 3 (September 2001).

Hughes, Sally Smith, 'Ronald Cape, Biotech pioneer and co-founder of Cetus', oral history conducted in 2003 by Sally Smith Hughes, Regional Oral History Office, The Bancroft Library, University of California-Berkeley (2006) pp. 541–75.

Hughes, Sally Smith, *Genentech, the Beginnings of Biotech* (Chicago: Chicago University Press, 2011).

Ingebretsen, Mark, *NASDAQ, a History of the Market that Changed the World* (Roseville, California: Prima Publishing, 2002).

Jones, Edgar, *The Business of Medicine, the Extraordinary History of Glaxo* (London: Profile Books, 2001).

Jones, Stephanie, *The Biotechnologists* (Basingstoke: Macmillan, 1992).

Kay, John, *The Kay Review of UK equity markets and long-term decision-making*, Final Report, Department of Trade and Industry, July 2012.

Kenney, Martin, *Biotechnology, the University–Industrial Complex* (New Haven, CT: Yale University Press, 1986).

Kenney, Martin, 'Schumpeterian innovation and entrepreneurs in capitalism: a case study of the US biotechnology industry', *Research Policy* 15 (1986) pp. 21–31.

Kettler, Hannah E. and Steven Casper, 'The Road to Sustainability in the UK and German Biotechnology Industries' (London: Office of Health Economics, 2000).

Kirk, Kate and Charles Cotton, *The Cambridge Phenomenon, 50 Years of Innovation and Enterprise* (London: Third Millennium Publishing, 2012).

Kneller, Robert, 'Autarkic drug discovery in Japanese pharmaceutical companies: insights into national differences in industrial innovation', *Research Policy* 32 (2003) pp. 1805–27.

Kneller, Robert, *Bridging Islands, Venture Companies and the Future of Japanese and American Industry* (Oxford: OUP, 2007).

Kneller, Robert, 'The importance of new companies for drug discovery: origins of a decade of new drugs', *Nature Reviews Drug Discovery* 9 (November 2010) pp. 867–82.

Kornberg, Arthur, *The Golden Helix, Inside Biotech Ventures* (Sausalito, CA: University Science Books, 1995).

Lacasa, Iciar Domingues, Thomas Reiss, and Jacqueline Senker, 'Trends and gaps in biotechnology policies in European member states since 1994', *Science and Public Policy* 31, 5 (October 2004) pp. 385–95.

Lambert, Richard, *Lambert review of business-industry collaboration*, final report (London: HM Treasury, December 2003).

Lazonick, William, 'Profits without prosperity: stock buybacks manipulate the market and leave most Americans worse off', *Harvard Business Review* (September 2014) pp. 3–11.

Lazonick, William and Öner Tulum, 'US biopharmaceutical finance and the sustainability of the biotech business model', *Research Policy* 40 (2011) pp. 1170–87.

Lea, William, 'Reorganisation of the Science Research Councils', House of Commons Library, Science and Environment Section, Research Paper 94/19 (House of Commons, 1 February 1994).

Lee, Vivien, John Gurnsey, and Arthur Klausner, 'New trends in financing biotechnology', *Nature Biotechnology* 1, 7 (September 1983) pp. 544–59.

Lerner, Josh, *Boulevard of Broken Dreams: Why Public Efforts to Boost Entrepreneurship and Venture Capital have Failed, and What to Do About It* (Princeton: Princeton University Press, 2009).

Lerner, Josh, Yannis Pierrakis, Liam Collins, and Albert Bravo Biosca, *Atlantic drift: venture capital performance in the UK and the US* (NESTA Research Report June 2011).

Levy, Jonah D. 'The return of the state? The economic policy of Nicolas Sarkozy', paper presented at the annual conference of the American Political Science Association, Washington DC, 2–5 September 2010.

Louët, Sabine, 'New law to boost French biotech industry', *Nature Biotechnology* 17 (1999) p. 1055.

Louet, Sabine, 'French genomics setup questioned', *Nature Biotechnology* 18 (2000) pp. 375–6.

Lundvall, B.-Å. (ed.) *National Innovation Systems: Towards a Theory of Innovation and Interactive Learning* (London: Pinter, 1992).

Lynn, Leonard H. and Reiko Kishida, 'Changing paradigms for Japanese technology policy: SMEs, universities and biotechnology', *Asian Business and Management* (2004) pp. 459–78.

Lynskey, Michael J., 'Bioentrepreneurship in Japan: institutional transformation and the growth of bioventures', *Journal of Commercial Biotechnology* 11, 1 (October 2004) pp. 9–37.

Malik, Khaleel, Luke Georghiou, and Hugh Cameron, 'Behavioural additionality of the UK SMART and LINK schemes', in *Government R & D funding and company behaviour, measuring behavioural additionality* (Paris: OECD, 2006).

Mandelson, Peter, *The Third Man* (London: HarperCollins, 2001).

Marks, Lara V., *The Lock and Key of Medicine: Monoclonal Antibodies and the Transformation of Healthcare* (London: Yale University Press, 2015).

Marks, Lara V, 'The life story of a biotechnology drug: Alemtuzumab', available on Lara Marks's website www.whatisbiotechnology.org.

Martin, Paul, Michael M. Hopkins, Paul Nightingale, and Alison Kraft, 'On a critical path: Genomics, the crisis of pharmaceutical productivity and the search for sustainability', in *The Handbook of Genetics and Society, Mapping the New Genomic Era*, edited by Paul Atkinson, Peter Glasner, and Margaret Lock (London: Routledge, 2009).

Martin, Paul, and Sandy Thomas, 'The "Commercialization Gap" in gene therapy: lessons for European competitiveness', in *Biotechnology and Competitive Advantage: Europe's Firms and the US Challenge*, edited by Jacqueline Senker, coordinated by Roland van Vliet (Cheltenham: Edward Elgar, 1998).

Maybeck, Vanessa and William Bains, 'Small company mergers—good for whom?' *Nature Biotechnology* 24 (2006), pp. 1343–8.

Mazzucato, Mariana, *The Entrepreneurial State, Debunking Private Sector vs. Public Sector Myths* (London: Anthem Press, 2013).

McKelvey, Maureen, *Evolutionary Innovations, the Business of Biotechnology* (Oxford: OUP, 1996).

McKelvey, Maureen, Anika Rickne, and Jens Laage-Hellman (eds), *The Economic Dynamics of Modern Biotechnology* (Cheltenham: Elgar, 2004).

Michie, Ranald, *The London Stock Exchange, a History* (Oxford: OUP, 1999).

Monopolies Commission, *A report on proposed mergers between Beecham and Glaxo and between Boots and Glaxo* (London: HC341, HMSO, July 1972).

Morange, Michel, *A History of Molecular Biology*, translated by Matthew Cobb (Cambridge, MA: Harvard University Press, 1998).

Mowery, David C., Richard R. Nelson, Bhaven N. Sampat, and Arvids A. Ziedonis, *Ivory Tower and Industrial Innovation, University-industry Technology Transfer before and after the Bayh–Dole Act* (Stanford, CA: Stanford University Press, 2004).

Mowery, David C. and Nathan Rosenberg, *Paths of Innovation, Technological Change in 20th Century America* (Cambridge: CUP, 1998).

Munos, Bernard, 'Lessons from 60 years of pharmaceutical innovation', *Nature Reviews Drug Discovery* 8 (2009) pp. 959–68.

Murmann, Johann Peter, *Knowledge and Competitive Advantage* (Cambridge: CUP, 2003).

Murmann, Johann Peter and Ralph Landau, 'On the making of competitive advantage: the development of the chemical industries in Britain and Germany since 1850', in *Chemicals and Long-term Economic Growth*, edited by Ashish Arora, Ralph Landau, and Nathan Rosenberg (New York: John Wiley, 1998).

Mustar, Philippe and Philippe Laredo, 'Innovation and research policy in France (1980–2000) or the disappearance of the Colbertist state', *Research Policy* 31 (2002) pp. 55–72.

Mustar, Philippe and Mike Wright, 'Convergence or path dependence in policies to foster the creation of university spin-off firms? A comparison of France and the United Kingdom', *Journal of Technology Transfer* 35, 1 (February 2010) pp. 42–65.

Myners, Paul, *A Study by Paul Myners into the Impact of Shareholders' Pre-emption Rights on a Public Company's Ability to Raise New Capital* (London: Department of Trade and Industry, February 2005).

Nelson, Richard R., 'The link between science and invention: the case of the transistor', in *The rate and direction of inventive activity: economic and social factors*, A report of the National Bureau of Economic Research, New York (Princeton: Princeton University Press, 1962).

Nelson, Richard R. (ed). *National Systems of Innovation, a Comparative Analysis* (Oxford: OUP, 1993).

Nelson, Richard R., 'What enables rapid economic progress: What are the needed institutions?' *Research Policy* 37 (2008) pp. 1–11.

Neven, Damien and Paul Seabright, 'European industrial policy: the Airbus case', *Economic Policy* 21 (1995) pp. 313–58.

Nightingale, Paul, Gordon Murray, Marc Cowling, Charles Baden-Fuller, Colin Mason, Josh Siepel, Michael M. Hopkins, and Charles Dannreuther, *From Funding Gaps to Thin Markets, UK government support for early-stage venture capital* (London: British Venture Capital Association and NESTA September 2009).

Oakley, Brian, and Kenneth Owen, *Alvey, Britain's Strategic Computing Initiative* (Cambridge, MA: MIT Press, 1989).

OECD Reviews of Innovation Policy: France (Paris: OECD, December 2014).

OECD Science, Technology and Industry Outlook (Paris: OECD, 2014).

Office for Life Sciences, *Life Sciences Blueprint* (Department for Business Innovation and Skills, July 2009).

Office for Life Sciences, *Strategy for UK Life Sciences* (Department for Business, Innovation and Skills, December 2011).

Office of Technology Assessment, *Commercial Biotechnology: An International Analysis* (Washington DC: US Congress, 1984).

Okamoto, Yumiko, 'Paradox of Japanese biotechnology: can the regional cluster development approach be a solution?' Department of Policy Studies, Doshisha University, Unpublished paper, December 2008.

O'Neill, Michael and Michael M. Hopkins, *A Biotech Handbook: A Practical Guide* (Oxford: Woodhead Publishing Series in Biomedicine, 2012).

Orsenigo, Luigi, *The Emergence of Biotechnology* (London: Pinter, 1989).

Owen, Geoffrey, *From Empire to Europe: The Decline and Revival of British Industry since the Second World War* (London: HarperCollins, 1999).

Owen, Geoffrey, 'Where are the big gorillas? High technology entrepreneurship in the UK and the role of public policy' (Diebold Institute for Public Policy Studies, December 2004) Available at cep.lse.ac/seminarpapers/09-05-05-OWE.pdf.

Owen, Nicholas, 'The pharmaceutical industry in the UK', in *Lessons from some of Britain's most successful sectors: an historical analysis of the role of government* (Department for Business, Innovation and Skills, Economics Paper No 6, March 2010).

Oxfordshire Economic Observatory, *Enterprising Oxford, the Growth of the Oxfordshire High-tech Economy* (Oxford, 2000).

Padgett, John F. and Walter W. Powell (eds), *The Emergence of Organisations and Markets* (Princeton: Princeton University Press, 2012).

Pammolli, F. L., Magazzini, and M. Riccaboni, 'The productivity crisis in pharmaceutical R&D', *Nature Reviews Drug Discovery* 10 (June 2011) pp. 428–38.

Patel, Pari, Antony Arundel, and Michael M. Hopkins, *Sectoral Innovation Systems in Europe: Monitoring, Analysing Trends and Identifying Challenges in Biotechnology* Final report, Europe Innova (European Commission 2008).

Perkins, Tom, *Valley Boy, the Education of Tom Perkins* (New York: Penguin Books, 2007).

Peters, Toine, *Interferon, the Science and Selling of a Miracle Drug* (Abingdon: Routledge, 2005).

Pisano, Gary P., *Science Business* (Boston: Harvard Business School Press, 2006).

Porter, Michael, 'Capital disadvantage: America's failing capital investment system', *Harvard Business Review* (September–October 1993).

Powell, Walter W., Kelley Packalen, and Kjersten Whittington, 'Organisational and institutional genesis: the emergence of high-tech clusters in the United States', in *The Emergence of Organisations and Markets*, edited by John F. Padgett and Walter W. Powell (Princeton: Princeton University Press, 2012).

Powell, Walter W. and Kurt Sandholtz, 'Chance, Nécessité et Naiveté, ingredients to create a new organisational form', in *The Emergence of Organisations and Markets*, edited by John F. Padgett and Walter W. Powell (Princeton: Princeton University Press, 2012).

Prevezer, Martha, 'Ingredients in the early development of the U.S. Biotechnology Industry', *Small Business Economics* 17 (2001) pp. 17–29.

Rabinow, Paul, *Making PCR: A Story of Biotechnology* (Chicago: University of Chicago Press, 2006).

Rasmussen, Nicolas, *Gene Jockeys, Life Science and the Rise of Biotech Enterprise* (Baltimore: John Hopkins University Press, 2014).

Reichert, J. M., C. J. Rosenweig, L. B. Faden, and M. C. Dewtiz, 'Monoclonal antibody successes in the clinic', *Nature Biotechnology* 23, 9 (2005) pp. 1073–8.

Reid, Graeme, *Why should the taxpayer fund science and research?* (National Centre for Universities and Business, December 2014).

Report of the Committee of Enquiry into the relationship of the pharmaceutical industry with the National Health Service (Cmnd 3410, HMSO, 1967).

Richards, Graham, *Spin-Outs, Creating Businesses from University Intellectual Property* (Petersfield: Harriman House, 2009).

Robbins-Roth, Cynthia, *From Alchemy to IPO, the Business of Biotechnology* (Cambridge, MA: Perseus Publishing, 2000).

Roe, Mark J., 'Corporate short-termism in the boardroom and in the courtroom', *Business Lawyer* 68, 4 (August 2013) pp. 977–1006.

Rothermael, Frank T., 'Incumbent's advantage through exploiting complementary assets via interfirm cooperation', *Strategic Management Journal* 22 (2001) pp. 687–99.

Rothman, H. and A. Kraft, 'Downstream and into deep biology: Evolving business models in "top tier" genomics companies', *Journal of Commercial Biotechnology* 12, 2 (January 2006), pp. 86–98.

Sainsbury, Lord, *Biotechnology clusters*, Report of a team led by Lord Sainsbury, Minister for Science (Department of Trade and Industry, August 1999).

Sainsbury, Lord, *The race to the top: a review of the government's science and innovation policies* (Department for Innovation, Universities and Skills, October 2007).

Saxenian, Annalee, *Regional Advantage* (Cambridge, MA: Harvard University Press, 1994).

Saxonhouse, Gary R., 'Industrial policy and factor markets: biotechnology in Japan and the United States', in *Japan's High-technology Industries: Lessons and Limitations of Industrial Policy*, edited by Hugh Patrick and Larry Meissner (Seattle: University of Washington Press, 1986).

Seabright, Paul, 'National and European champions—burden or blessing?' *CESifo Forum* 6, 2 (Ifo Institute for Economic Research, Munich, Summer 2005).

Segal Quince Wicksteed, *The Cambridge Phenomenon Revisited* (Cambridge, June 2000).

Senker, Jacqueline, 'Biotechnology: the external environment', in *Biotechnology and Competitive Advantage: Europe's Firms and the US Challenge*, edited by Jacqueline Senker, co-ordinated by Roland van Vliet (Cheltenham: Edward Elgar, 1998).

Sharp, Margaret, *The New Biotechnology: European Governments in Search of a Strategy*. Industrial Adjustment and Policy: VI Sussex European Papers No.15. Science Policy Research Unit, University of Sussex (1985).

Sharp, Margaret, 'Biotechnology in Britain and France: the evolution of policy', in *Strategies for New Technology, Case Studies from Britain and France*, edited by Margaret Sharp and Peter Holmes (Hemel Hempstead: Philip Allan, 1989).

Sharp, Margaret, 'The management and coordination of biotechnology in the UK 1980–1988', *Philosophical Transactions of the Royal Society of London, Series B, Biological sciences* 324, 1224 (31 August 1989).

Sharp, Margaret, 'Biotechnology and advanced technology infrastructure policies: the example of the UK's protein engineering club', *Economics of Science, Technology and Innovation* 7 (1996) pp. 217–46.

Sheridan, Cormac, 'Cimzia's setback paves way for other TNF inhibitors in Crohn's disease', *Nature Biotechnology* 25 (2007) pp. 487–8.

Shire plc, *Flying high, the first twenty years at Shire Pharmaceuticals* (Shire plc, 2006).

Siepel, Josh, 'Capabilities, policy and institutions in the emergence of venture capital in the UK and the US', D Phil thesis, University of Sussex, Brighton (2009).

Smith, Graham, Muhammad Safwan Akram, Keith Redpath, and William Bains, 'Wasting cash—the decline of the British biotech sector', *Nature Biotechnology* 27 (2009) pp. 531–7.

Tansey, E. M. and P. P. Catterall, 'Technology transfer in Britain: The case of monoclonal antibodies', *Contemporary Record* 9, 2 (1995) pp. 409–44.

Teitelman, Robert, *Gene Dreams: Wall Street, Academia, and the Rise of Biotechnology* (New York: Basic Books, 1989).

Temin, Peter, 'Technology, regulation and market structure in the modern pharmaceutical industry', *Bell Journal of Economics* 10 (1979), pp. 429–46.

Thomas, Lacy Glenn, 'Implicit industrial policy: the triumph of Britain and the failure of France in global pharmaceuticals', *Industrial and Corporate Change* 3, 2 (1994) pp. 451–89.

Thomas, Lacy Glenn, *The Japanese Pharmaceutical Industry, the New Drug Lag and the Failure of Industrial Policy* (Cheltenham: Elgar, 2001).

Tilton, John, *International Diffusion of Technology: The Case of Semiconductors* (Washington DC: Brookings Institution, 1971).

UK Government, *Biotechnology* (Cmnd 8177, HMSO, March 1981).

UK Government, 'Realising our potential, a strategy for science, engineering and technology', presented to Parliament by the Chancellor of the Duchy of Lancaster (Cmnd 2250, HMSO, May 1993).

Umemura, Maki, 'Crisis and change in the system of innovation: the Japanese pharmaceutical industry during the Lost Decades, 1990–2010', *Business History*, 56, 5 (2014) pp. 816–44.

Vallas, Steven P., Daniel Lee Kleinman, and Dina Biscotti, 'Political structures and the making of US biotechnology, in *State of Innovation, the US Government's Role in Technology Development*, edited by Fred Block and Matthew R. Keller (Boulder: Paradigm Publishers, 2011).

Vettel, Eric J., *Biotech: The Counter-cultural Origins of an Industry* (Philadelphia: University of Pennsylvania Press, 2006).

Walker, William, 'National innovation systems: Britain', in *National Innovation Systems, a Comparative Analysis*, edited by Richard R. Nelson (Oxford/New York: OUP, 1993).

Watson, James D., *The Double Helix* (London: Weidenfeld and Nicolson, 1968).

Werth, Barry, *Billion Dollar Molecule, One Company's Search for the Perfect Drug* (New York: Simon & Schuster, 1994).

Werth, Barry, *The Antidote, Inside the World of New Pharma* (New York: Simon & Schuster, 2014).

Whittington, Kjersten Bunker, Jason Owen-Smith, and Walter W. Powell, 'Networks, propinquity and innovation in knowledge-intensive industries', *Administrative Science Quarterly* 54, 1 (March 2009) pp. 90–122.

Wieland, Thomas, 'Ramifications of the "Hoechst shock": perceptions and cultures of molecular biology in Germany', Working Paper, Munich Centre for the History of Science and technology (August 2007).

Willetts, David, *Eight great technologies* (London: Policy Exchange, 2013).

Willott W. B. 'The NEB involvement in electronics and information technology', in *Industrial Policy and Innovation*, edited by Charles Carter (London: Heinemann, 1981).

Wood, Andrew, 'Direct costs of share issues on public markets—reviving the debate over non-pre-emptive share issues' (BioIndustry Association, 16 December 2012).

Yi, Doogab, 'Who owns what? Private ownership and the public interest in Recombinant DNA Technology in the 1970s', *Isis* 102, 3 (September 2011) pp. 446–74.

Yi, Doogab, *The Recombinant University: Genetic Engineering and the Emergence of Stanford Biotechnology* (Chicago: Chicago University Press, 2015).

Zucker, Lyman G., Michael R. Darby, and Marylyn B. Brewer, 'Intellectual human capital and the birth of US biotechnology enterprises', *American Economic Review* 88, 1 (March 1998) pp. 290–306.

Index

Note: Country of origin is given in brackets for pharma/biotech firms other than those based in the UK. Figures and tables are indicated by an italic *f* and *t* following the page number.

Index

Index